Edward L. Wilson

The production of photographic prints in permanent pigments

Edward L. Wilson

The production of photographic prints in permanent pigments

ISBN/EAN: 9783742860316

Manufactured in Europe, USA, Canada, Australia, Japa

Cover: Foto ©berggeist007 / pixelio.de

Manufactured and distributed by brebook publishing software (www.brebook.com)

Edward L. Wilson

The production of photographic prints in permanent pigments

THE

AMERICAN CARBON MANUAL:

OR,

THE PRODUCTION OF PHOTOGRAPHIC PRINTS IN PERMANENT PIGMENTS.

BY EDWARD L. WILSON,

Editor of The Philadelphia Photographer.

NEW YORK:
SCOVILL MANUFACTURING COMPANY.
1868.

PREFACE.

One of the greatest hindrances to the progress of the art of Photography, is the doubtful permanency of its productions. It will become an immense power if we can overcome this objection. Its votaries have learned how to secure beautiful lights and shades; brilliant chemical effects and most artistic and pleasing pictures. Never before has our art shown greater promise of improvement; never before has there been such a thirst for information and thorough training exercised as at present; and yet what a mortification to every earnest photographer to feel that his efforts can produce nothing that will bear him honor and credit longer than a few short years!

With this in view, constantly staring us in the face, is it not strange that the growth of photography has been so great as it has, and its improvement so evident as it is?

It would be so, were it not that one ray of hope has beckoned us on for a number of years back, *i. e.*, the hope that at no distant period we might produce permanent results. That hope is now fully realized in the Carbon printing process, several of which are described herein, by which we may produce permanent photographic prints.

Many, however, are debarred from its practice by the want of a proper practical manual of instruction.

The object of this work is to observe, collate, and condense, as far as possible, the best and most practical thoughts of the few who have experimented with, and written on this process, not only abroad but at home, and to combine them with my own experience and observation.

I am particularly indebted to my friend G. Wharton Simpson, A.M., editor of the "*Photographic News*," London, whose excellent manual "On the Production of Photographs in Pigments—Swan's Process"—has recently been given to the world, and who is one of the earliest experimenters in the process. I also gratefully acknowledge the receipt of some useful and practical ideas from Dr. Herman Vogel, editor of the *Photo. Mittheilungen*, Berlin, translator and author of a revise of Mr. Simpson's work.

By careful and extended experiment I am able to indorse their views, and to add a few notes here and there of what has occurred to me in my own practice, and several other matters which I trust will be found useful. It is more difficult to sift out from a large amount of thought and record that which is most important to know—using one's brains as a sieve or filter to separate the good grain from the tares—the sediment from the pure solution, as it were—than to scatter the seed—to distribute the mixture—as one has gathered it from one's own experience and experiment. I have endeavored, however, to compile and write such matter for these pages as will make the novel, fascinating and valuable Carbon Process, plain, practical and easy to all workers in photography.

Although it will be seen by the historical notes that follow, that many carbon processes have been worked more or less, yet to Mr. Joseph Wilson Swan, New Castle-upon-Tyne, England, we are indebted for the most practical and perfect one, and to this I will ask you to give your especial attention. It is now practised considerably and successfully in this country, as our specimen will testify, and most largely abroad, by Messrs. Swan, Braun, and others.

Like many other useful discoveries, this process has been perfected only by slow degrees, and by the laborious and patient research of many individuals. As early as 1814 M. Niepce made experiments in Carbon printing, and to him it owes its origin; but after all the manifold experiments by the many who shall be named in proper place, to Mr. Swan is due the honor and praise for having simplified, perfected, and made easy of practice, a process for photographic printing, where every beauty of management and manipulation is preserved perfectly and permanently.

Those who endeavor to practise it will find it entirely different from the silver process—no gold, no silver, no hyposulphite, entering the construction of a carbon print—yet quite as easy and possessed of many advantages. If what follows should not meet every case, any who may be troubled with failures I shall be glad to answer through the columns of the *Philadelphia Photographer*, wherein also notes of future experiments, improvements and discoveries in the process shall be regularly recorded.

EDWARD L. WILSON.

PHILADELPHIA, May, 1868.

CONTENTS.

THE

AMERICAN CARBON MANUAL:

OR, THE PRODUCTION OF PHOTOGRAPHS IN PERMANENT PIGMENTS.

INTRODUCTION.

A CARBON PHOTOGRAPH, in the strict sense of the word, is an image in carbon produced by the action of light. The term, as commonly used by photographers, has, however, a wider application, and is employed to designate any sun picture produced in permanent pigments, whether consisting of carbon or not.

Almost all the methods which have been proposed for the production of such pictures depend upon one principle. They are based on the fact that light renders certain soluble bodies insoluble in the usual menstrua. This principle admits of varied application in producing pictures; but in the processes which have been brought to the highest practical perfection, some coloring matter—Indian ink or lampblack for instance—has been added to a colorless body like gelatine or gum, which, on being rendered insoluble by the action of light in parts, imprisons the coloring matter, and thus forms the dark parts of the image. A sheet of paper coated with such a substance mixed with a pigment, exposed to light under a negative, and then washed to remove all the soluble matter, will produce a picture, the blacks of which are formed by the insoluble substance and pigment, and the whites by the surface of the paper from which the colored coating has been washed away.

It will be seen, however, that in the production of an image by means of a photographic negative, in coloring matter so imprisoned, there is no provision for the rendering of gradation of tint. A layer of substance capable of being rendered insoluble by the action of light, if extended on a sheet of paper and exposed to light under a stencil plate, would be rendered insoluble wherever the luminous action penetrated the apertures in the plate. If the paper were then treated with a solvent of the substance with which it was coated, the coating would be removed from all portions protected from the action of light by the opaque parts of the plate, and a perfect transcript of the design would be formed on a white ground. If, instead of the stencil plate, a photographic negative be employed, the image in which is formed by varying gradations of opacity, the result is somewhat different. The layer of soluble matter is rendered insoluble wherever the light has penetrated sufficiently through the transparent parts of the negative; but where the more opaque parts of the negative, through which light has penetrated with much less intensity, protect the coated surface, a portion only of the coating is rendered insoluble, that portion being the surface in immediate contact with the negative. When the prepared paper is submitted to the action of a solvent, the thoroughly-exposed portions, being quite insoluble, are not removed, but those parts representing the lighter tones of the picture, having become insoluble on the upper surface only, the layer underneath is readily dissolved, and the whole film in such parts is thus removed by the solvent. An imperfect image, possessing only deep blacks and masses of white without gradation of half-tone, is the result.

This was the great difficulty of carbon printing in the early experiments which succeeded the discovery by M. Poiteven, in 1855, of the principles upon which it is based, although the cause was not at first fully understood. After two or three years of comparatively unsuccessful effort, it was discovered, that in order to succeed in producing gradations of half-tone in such pictures, it was necessary to wash away the unaltered and still soluble matter at the side of the film opposite to that exposed to light, in order to preserve intact every portion of the film which had been rendered insoluble, and so leave a film varying in thickness according to the depth to which light had penetrated; this depth being governed by the varying degrees of transparency of the different parts of the negative. This varying thickness of a colored translucent film upon a white ground, accordingly rendered the gradations of the picture, a thick layer representing deep blacks, and thinner layers, various gradations of half-tint.

This principle recognized, it became possible to produce carbon prints, which were true transcripts of a photographic negative; but a mode of rendering the principle practically useful was wanting. The first idea was to place the back of the prepared paper in contact with the negative, so that the light should, after traversing the paper, act upon that surface of the sensitive layer which was in contact with the paper, rendering it insoluble in varying degrees. The soluble portion was then washed away from the upper surface, leaving undisturbed the insoluble layer of various thicknesses, which forms the picture on the paper. This method was found to be comparatively impracticable from the long exposure it rendered necessary; the picture suffered also, and became coarse, from the granular texture of the paper through which the light had to pass. Various methods were tried, the details of which appear in a subsequent chapter; but none were found efficient until Mr. Swan discovered the method of preparing a sensitive film, which, having been exposed to light in direct contact with the negative, could be transferred to another support, so as to permit the easy washing away of the unaltered material at the side opposite to that which had been exposed to light. This method was found not only more simple in practice, but more excellent in result than any method previously attempted. It gave pictures of exquisite delicacy and force, rendering perfectly every gradation in the negative with a degree of beauty which had rarely been obtained even by the usual methods of silver printing.

It is unnecessary to enter into a consideration here of the possibility of producing permanent pictures by the usual photographic processes of silver printing. The stigma of instability has been hitherto the chief drawback to the beautiful results of photography. In no silver printing process has immunity from fading been obtained; and although many photographs have been preserved unchanged for years, others, produced under apparently the same conditions, have become, during the same period, faded and worthless; and uncertainty of permanence, if not certainty of fading, remains a characteristic of all silver prints.

In Swan's carbon process the image is produced in the pigments of the painter, and whatever of permanence may be predicated of a water-color drawing, may be affirmed of the photographs produced in the same materials. If carbon be regarded as the most stable coloring matter which can be employed, lampblack or other similar preparations of carbon can be used in the production of the print. Sepia, bistre, Indian-ink, or any combination of these or other pigments, by which a pleasing monochrome may be produced and permanency attained, is equally available in this method of printing.

Thus, a more certain control over the tone and quality of the picture is obtained, whilst absolute stability is secured. The surface of pictures produced by this process is devoid of the gloss, which is regarded by many as vulgar, in prints on albumenized paper; the lights being absolutely flat or dead, the shadows only presenting, in some cases, a slightly glossy surface. It will be thus seen that the pictures gain not less in artistic qualities of color and texture than in permanency.

Mr. Adolph Braun, Dornach (Haut Rhin) France, has already applied the carbon process successfully in a direction heretofore considered almost impossible to reach. He has made negatives of thousands of the drawings of the old masters in the art museums of Europe, and reproduced them in the colors of the originals, which embrace almost every shade and color.

This is an immense reach for photography, and will greatly lead towards the elevation of the art, and the cultivation of artistic taste. Mr. Braun has purchased Mr. Swan's patent for France and Belgium, and is the largest worker in carbon printing in the world.

It is not necessary, however, to enter into more extended comment here on the beauty of the results, as the reader will form his own opinion from an examination of the print which accompanies these pages.

Before concluding these introductory remarks, it may

be desirable to give a brief summary of the grounds for stating that permanency is secured in the method of printing to be described: They are these: the image is formed of carbon or other known permanent pigment; the vehicle or menstruum in which the pigment is held is gelatine, rendered quite insoluble by combination with the oxide of chromium; the amount of gelatine present in the image, and uniting it with the paper on which it rests, is not greater than that on the surface of a sheet of well-sized writing-paper, so that there is as little danger of the vehicle cracking or decomposing, as there is of the pigment fading. Absolute imperishability cannot probably be predicated of any picture formed of pigments and paper; but, as much permanency may be anticipated for these prints as is found to pertain to an Indian-ink drawing, to which experience permits us to award a duration of at least several centuries.

SWAN'S CARBON PROCESS.

PRIOR to entering into a detailed description of Mr. Swan's Carbon Process, it may be desirable to place before the reader the specification of the patent in which is stated the essential features of principle and practice combined in the process. It proceeds as follows:

"My invention relates to that manner or style of photographic printing known as carbon or pigment printing. In this style of printing, carbon or other coloring matter is fixed by the action of light passing through a negative, and impinging upon a surface composed of gelatine, or other like substance, colored with carbon or other coloring matter, and made sensitive to light by means of bichromate of potash, or bichromate of ammonia, or other chemical substance having like photographic property; those portions of the colored and sensitive gelatinous surface which are protected from the light by the opaque or semi-opaque portions of the negative, being afterwards washed away by means of water, while the parts made insoluble by light remain, and form a print. This kind of photographic printing, although possessing the advantage of permanency, and affording the means of insuring any required tone or color for the print, has not come into general use, because of the difficulties hitherto experienced in obtaining by it delicacy of detail, and complete gradation of light and shade.

"The difficulties referred to were more particularly experienced in attempts to employ paper coated with the colored gelatinous materials, and arose from the fact, that, in order to obtain half-tone, certain portions of the colored coating lying behind or at the back of the photographically-impressed portions required to be washed away, and the employment of paper in the way it has been employed hitherto, not only as a means of supporting the colored coating, but also to form ultimately the basis or groundwork of the print, obstructed the removal of the inner or back portions of the colored coating, and prevented the obtaining of half-tone.

"Now, my invention consists in the formation of tissues adapted to the manner of printing referred to, and composed of, or prepared with, colored gelatinous matter, and so constructed, that while they allow, in the act of printing, free access of light to one surface of the colored gelatinous matter, they also allow free access of water, and the unobstructed removal of the non-affected portions of the colored matter, from the opposite surface, or back, in the act of developing; and I obtain this result either by the disuse of paper altogether, or by the use of it merely as a backing or temporary support of the colored gelatinous matter; the paper, so used, becoming entirely detached from the colored gelatinous matter in the act of developing, and forming no part of the print ultimately.

"My invention consists, furthermore, in the special mode of using the said tissues, whereby superior half-tone and definition in the print are obtained as aforesaid, and also in a mode of transferring the print, after developing, from a temporary to a permanent support, so as to obtain a correction in the position of the print in respect of right and left. In producing the photographic tissues referred to, I form a solution of gelatine, and for the purpose of imparting pliancy to the resultant tissue, I have found it advisable to add to the gelatine solution, sugar or other saccharine matter, or glycerine. To the said gelatinous solution I add carbonaceous or other coloring matter, either in a fine state of division, such as is used in water-color painting, or in the state of a solution or dye, or partly in a fine state of division, and partly in solution.

"With this colored gelatinous solution I form sheets or films, as hereafter described; and I render such sheets or films sensitive to light, either at the time of their formation, by introducing into the gelatinous compound bichromate of ammonia, or other agent of like photographic properties, or by applying to such non-sensitive sheets or films, after their formation, a solution of the bichromate, or other substances of like photographic property. This latter method I adopt when the sheet or film is not required for use immediately after its formation. I will, in my future references to the bichromate of ammonia or the bichromate of potash, or to other chemicals possessing analogous photographic properties, denominate them '*the sensitizer;*' and in referring to the colored gelatinous solution, I will denominate this mixture '*the tissue-compound.*' When the tissue to be produced is required for immediate use, I add the sensitizer to the tissue compound; but, where the tissue is required to be preserved for some time before using, I prefer to omit the sensitizer from the tissue-compound, with a view to the tissue being made sensitive to light subsequently, by the application of a solution of the sensitizer.

"With respect to the composition of the tissue-compound, it will be understood by chemists, that it may be varied without materially affecting the result, by the addition or substitution of other organic matters, similarly acted upon by light, when combined with a salt of chromium, such as I have referred to. Such other organic matters are gum arabic, albumen, dextrine; and one or more of these may be employed occasionally to modify the character of the tissue-compound, but I generally prefer to make it as follows: I dissolve, by the aid of heat, two parts of gelatine, in eight parts of water, and to this solution I add one part of sugar, and as much coloring matter in a finely divided state, or in a state of solution, or both, as may be required for the production of a photographic print with a proper gradation of light and shade. The quantity required for this purpose must be regulated by the nature of the coloring matter employed, and also by the character of the negative to be used in the printing operation. Where it is desired that the coloring matter of the print should consist entirely or chiefly of carbon, I prefer to use lampblack finely ground and prepared as for water-color painting, or I use Indian-ink; and where it is desired to modify the black, I add other coloring matter to produce the color desired. For instance, I obtain a purple black by adding to the carbon, indigo and crimson-lake, or I add to the carbon an aniline dye of a suitable color; where the coloring matter used is not a solution or dye, but solid matter in a fine state of division, such as Indian-ink or lampblack, I diffuse such coloring matter through water, or other inert liquid capable of holding it in suspension; and after allowing the coarser particles to subside, I add, of that portion which is held in suspension, as much as is required, to the gelatine solution. In preparing tissue to be used in printing from negatives technically known as 'weak,' I increase the proportion of coloring matter relatively to that of the tissue-compound; and I diminish it, for tissue or paper to be used in printing from negatives of an opposite character.

"Having prepared the tissue-compound as before described, I proceed to use it as follows: For preparing sensitive tissue, I add to the tissue-compound more or less of the sensitizer, varying the quantity added, according to the nature of the sensitizer, and to the degree of sensitiveness to be conferred on the tissue to be produced from it. For ordinary purposes, and where the tissue-compound is made according to the formula before given, I add about one part of a saturated solution of bichromate of ammonia to ten parts of the tissue-compound; and I make this addition immediately previous to the preparation of the tissue, and I maintain the tissue-compound in the fluid state, by means of heat, during the preparation of the tissue, avoiding the use of an unnecessary degree of heat; I also filter it through fine muslin or flannel, or other suitable filtering medium, previous to use; and I perform all the operations with the tissue-compound, subsequent to the introduction of the sensitizer, in a place suitably illuminated with yellow or non-actinic light. In forming tissue upon a surface of glass, I first prepare the glass, so as to facilitate the separation of the tissue from it. For this purpose, I apply ox-gall to the surface of the glass (by means of a brush, or by immersion), and allow it to dry. The glass is then ready for coating with the tissue-compound, or I apply to the glass a coating of collodion, previous to the application of the coating of tissue-compound. In this case, the preparation with ox-gall is unnecessary. When collodion is used, the collodion may consist of about ten grains of pyroxyline in one ounce of mixture of equal parts of sulphuric ether and alcohol. I apply the collodion by pouring it on the surface to be coated, and draining off the excess, and I allow the coating of collodion to become dry before applying the coating of tissue-compound. I generally use a plane surface on which to form the tissue, but surfaces of a cylindrical or other form may sometimes be used advantageously. In preparing sheets of sensitive tissue on a plane surface of glass, I prefer to use the kind of glass known as plate, or patent plate. Before applying the sensitive tissue-compound, I set the plate to be coated, so that its upper surface lies in a horizontal position, and I heat the plate to about the same temperature as the tissue-compound, that is, generally, to about 100 degrees, Fahrenheit. The quantity of the tissue-compound that I apply to the glass varies with circumstances, but is generally about two ounces to each square foot of surface coated.

After pouring the requisite quantity of the tissue-compound upon the surface of the plate, I spread or lead the fluid by means of a glass rod or soft brush, over the entire surface, taking care to avoid the formation of air-bubbles; and I keep the surface in the horizontal position, until the solidification of the tissue-compound. In coating other than plane surfaces, I vary, in a suitable manner, the mode of applying the tissue-compound to such surfaces. In coating a cylindrical surface, I rotate the cylinder in a trough containing the tissue-compound, and after having produced a uniform coating, I remove the trough, and keep up a slow and regular rotation of the cylinder until the coating has solidified. After coating the surface of glass or other substance as described, I place it in a suitable position for rapid drying, and I accelerate this process by artificial means, such as causing a current of dry air to pass over the surface coated, or I use heat, in addition to the current of air, or I place it in a chamber containing quick-lime, chloride of calcium, or other substance of analogous desiccating property. When the tissue is dry, I separate it from the surface on which it was formed, by making an incision through the coating to the glass, around the margin of the sheet; or I cut through the cylindrical coating near the ends of the cylinder, and also cut the coating across, parallel with the axis of the cylinder, when, by lifting one corner, the whole will easily separate in a sheet. When the tissue-compound is applied over a coating of collodion, the film, produced by the collodion, and that produced by the tissue-compound, cohere, and the two films form one sheet. Sometimes, before the separation of the coating from the glass, I attach to the coating a sheet of paper, for the purpose of strengthening the tissue, and making it more easy to manipulate. I generally apply the paper, in a wet state, to the dry gelatinous surface; and having attached the paper thereto in this manner, I allow it to dry; and I then detach the film and adherent paper from the glass, by cutting around the margin of the sheet, and lifting it off as before described. Where extreme smoothness of surface, such as is produced by moulding the tissue on glass, as described, is not of importance; and where greater facility of operation is desired, I apply a thick coating of the tissue-compound to the surface of a sheet of paper. In this case, the paper is merely used as a means of forming, and supporting temporarily, the film produced from the tissue-compound; and such paper separates from the gelatinous coating in a subsequent stage of my process. In coating a surface of paper with the sensitive tissue-compound, I apply the sheet, sometimes of considerable length, to the surface of the tissue-compound contained in a trough, and kept fluid by means of heat, and I draw or raise the sheet or length of paper off the surface with a regular motion; and I sometimes apply more than one coating to the same sheet in this manner. After such coating, I place the coated paper where it will quickly dry, and seclude it from injurious light.

"The sensitive tissue, prepared as before described, is, when dry, ready to receive the photographic impression, by exposure under a negative in the usual manner, or by exposure in a camera obscura, to light transmitted through a negative in the manner usual in printing by means of a camera. I prefer to use the sensitive tissue within two days of the time of its preparation. Where the tissue is not required for immediate use, I omit the sensitizer from the tissue-compound as before mentioned; and with this non-sensitive tissue-compound, I coat paper, glass, or other surface, as described in the preparation of the sensitive tissue or paper. In preparing sheets of non-sensitive tissue by means of glass, as described, I use no preliminary coating of collodion. I dry the non-sensitive tissue in the same manner as the sensitive, except that in the case of the non-sensitive tissue, seclusion from daylight is not necessary.

"The non-sensitive tissue is made sensitive, when required for use, by floating the gelatinous surface upon a solution of the sensitizer, and the sensitizer that I prefer to use for this purpose is an aqueous solution of the bichromate of potash containing about two and a half per cent. of this salt. I apply the sensitizer (by floating or otherwise), to the gelatinous surface of the tissue; and after this, I place it in a suitable position for drying, and exclude it from injurious light.

"In applying to photographic printing the various modifications of the sensitive tissue, prepared as before described, I place the sensitive tissue on a negative in an ordinary photographic printing-frame, and expose to light in the manner usual in photographic printing; or I place it in a camera obscura in the manner usual in printing by means of a camera obscura. When the tissue employed is coated with a film of collodion on one side, I place the collodionized side in contact with the negative; or where it is used in the camera, I place the collodionized side towards the light passing through the camera lens. Where the tissue is not coated with collodion, and where paper forms one of the surfaces of the tissue, the other surface being formed of a coating or film of the tissue-compound, I place this last-named surface in contact with the negative; or, when using it in the camera, I present this surface towards the light transmitted by the lens. After exposure for the requisite time, I take the tissue from the printing-frame or camera, and mount it in the manner hereinafter described, that is to say, I cement the tissue, with its exposed surface, or, in other words, with that surface which has received the photographic impression, downward, upon some surface (usually of paper) to serve temporarily as a support during the subsequent operation of developing, and with a view to the transfer of the print, after development, to another surface; or I cement it (also with the exposed or photographically impressed surface downward), upon the surface to which it is to remain permanently attached. The surface, on which it is so mounted, may be paper, card, glass, porcelain, enamel, &c. Where the tissue has not been coated with collodion previous to exposure to light, I prefer to coat it with collodion on the exposed or photographically impressed side, before mounting it for development, but this is not absolutely necessary; and I sometimes omit the coating with collodion, more particularly where the print is intended to be colored subsequently. Or where I employ collodion, with a view to connect the minute and isolated points of the print firmly together during development, I sometimes ultimately remove the film it forms, by means of a mixture of ether and alcohol, after the picture has been finally mounted, and the support of the film of collodion is no longer required. In mounting the exposed tissue or paper previous to development, in the temporary manner, with a view to subsequent transfer to another surface, I employ, in the mounting, a cement that is insoluble in the water used in the developing operation, but that can be dissolved afterwards, by the application of a suitable solvent; or one that possesses so little tenacity, that the paper or other support, attached temporarily to the tissue or paper by its means, may be subsequently detached without the use of a solvent.

"The cements that may be used for temporary mounting are very various, but I generally prefer to use a solution of India-rubber, in benzole or other solvent, containing about six grains of India-rubber in each ounce of the solvent, and I sometimes add to the India-rubber solution a small proportion of dammar-gum, or gutta-percha. In using this cement, I float the photographically impressed surface of the tissue upon it, and I treat, in a similar manner, the paper or other surface intended to be used as the temporary mount or support during development; and, after allowing the benzole or other solvent to evaporate, and while the surfaces coated with the cement are still tacky, I press them strongly together in such a manner as to cause them to cohere.

"When the photographically impressed, but still undeveloped tissue is to be cemented to a surface, that not only serves to support the picture during its development, but also constitutes permanently the basis of the picture, I prefer to use albumen or starch paste as the cementing medium; and where I employ albumen I coagulate or render it insoluble in water (by means of heat, by alcohol, or other means), after performing the cementing operation, and previous to developing. In the permanent, as in the temporary mode of mounting, I cement the tissue, *with its photographically impressed surface downwards*, upon the surface to which it is to be permanently attached. After mounting the tissue, as before described, and allowing the cement used time to dry, where it is of such a nature as to require it, I then submit the mounted tissue to the action of water, sufficiently heated to cause the solution and removal of those portions of the colored gelatinous matter of the tissue which have not been rendered insoluble by the action of light during exposure in the printing-frame or camera. Where paper has been used as a part of the original tissue; this paper soon becomes detached by the action of the warm water, which then has free access to the under stratum or back of the colored gelatinous coating, and the soluble portions of it are therefore readily removed by the action of the water; and by this means the impression is developed that was produced by the action of light during the exposure of the tissue in the printing-frame or camera, and the picture remains attached to the mount, cemented to the photographichlly impressed surface previous to development. I allow the water to act upon the prints during several hours, so as to dissolve out the decomposed bichromate as far as possible. I then remove them from the water, and allow them to dry, and those not intended for transfer, but that have been permanently attached to paper, previous to development, I finish by pressing and trimming in the usual manner. Those which have been temporarily mounted, I transfer to paper, card, or other surface. In transferring to paper or card, I coat the surface of the print with gelatine, gum arabic, or other cement of similar character, and allow it to dry. I then trim the print to the proper shape and size, and place its surface in contact with the piece of paper or card to which the transfer is to be effected, such piece of paper or card having been previously moistened with water, and I press the print and mount strongly together; and, after the paper or card has become perfectly dry, I remove the paper or other supporting material, temporarily attached, previous to development, either by simply tearing it off, where the cement used in the temporary mounting is of a nature to allow of this without injury to the print, or I apply to the temporary mount, benzole or turpentine, or other solvent of the cement employed, or I immerse the print in such solvent, and then detach the temporary mount, and so expose the reverse surface of the print; and, after removing from the surface of the print, by means of a suitable solvent, any remains of the cement used in the temporary mounting, I finish the print by pressing in the usual manner. If, however, the print be collodionized, and be required to be tinted with water-color, I prefer to remove the collodion film from the surface of the print, and this I do by the application of ether and alcohol.

"Having now set forth the nature of my invention of 'Improvements in Photography,' and explained the manner of carrying the same into effect, I wish it to be understood, that under the above in part recited letters-patent, I claim: First, the preparation and use of colored gelatinous tissues in the manner and for the purpose above described.

"Secondly, the mounting of undeveloped prints, obtained by the use of colored gelatinous tissues, in the manner and for the purpose above described.

"Thirdly, the re-transfer of developed prints, produced, as above described, from a temporary to a permanent support."

It will be seen that one of the essential features of the process is the production of a "tissue," which renders the manipulations necessary to perfect results practical and easy, which were before difficult or impossible. We shall see in the historical notes which follow that the principles upon which carbon printing is based had received partial recognition at an early period. As they became more perfectly understood the practical difficulties seemed greater. The imperative condition upon which half-tone depends, the exposure of one side of the film to light, and of the other to the solvents which should remove the unaltered material, seemed to present an insuperable difficulty. Exposure *through* the prepared paper was attended by two grave difficulties: the passage of the light through the paper, rendered yellow by saturation with the bichromate, was exceedingly protracted; and actinic force transmitted by the negative was to a great extent arrested before it reached the sensitive coating of gelatine and pigment, a brown tint (highly adiactinic) being formed in the texture of the paper. The intervening paper between the negative and sensitive layer is further objectionable on account of the loss of brilliancy incidental to its becoming more deeply brown where the shades of the print come, and consequently offering a proportionally greater obstruction to the light there than elsewhere. Besides, the finished picture, no matter how delicate the negative, bore marks of all the granulation or defects in texture of the paper through which the light passed. The ingenious device of M. Fargier, in which the sensitive coating was applied to a plate of glass, and, after exposure, transported from the glass, by the aid of a film of collodion, in order to wash away or develop on the opposite side of the film, was manifestly impractical on a commercial scale, especially for large pictures, owing to the difficulty of manipulating a thin film of gelatine attached to a thin film of collodion, floating in a vessel of water; and the difficulty, almost amounting to impossibility, of transferring such a film perfectly to paper. The "tissue" described in Mr. Swan's specification, together with the series of ingenious devices accompanying its use, have been the means of effectually overcoming the many practical difficulties above enumerated, and of making the production of carbon prints, of any size, in any number, and of unimpeachable quality, a thing commercially practicable.

There are two modes of proceeding in forming the tissue, either of which may be adopted, as circumstances may render desirable. The first consists in coating a plate of glass with plain collodion, and upon this film applying a mixture of gelatine, sugar, coloring matter, and bichromate of potash. When this is dry, it is removed from the glass, and forms a pliant tissue, ready for exposure under a negative, after which it is mounted (exposed face down) on paper, either temporarily, by means of caoutchouc cement, or permanently, by means of albumen. It is then developed, and, if temporarily mounted, is re-transferred, as we shall describe. The second and usual method consists in the application of the gelatine and coloring matter to paper, which can be rendered sensitive, when required, by immersion in a solution of bichromate of potash. It is then exposed under a negative, attached to another temporary basis of paper by a waterproof cement, the first paper being readily removed by soaking in warm water, so as to expose to the water the side of the film opposite to that which was in contact with the negative, the development and transference following in due course. We shall describe both these methods in detail.

A patent has been issued to John C. Crosman, Boston, Mass., for an "*Improved Process of Coating Sheets of Paper and Other Material with Solutions.*" His specifications read as follows:

"My invention relates to a process by which sheets of various material, such as leather, cloth, paper, &c., are covered by a coating applied in the form of a fluid, or a fluid solution, in such a manner that the resulting coating will be smooth and of uniform thickness, and so that when the solution applied contains chemical salts these will be equally distributed over the surface which is so covered. Said process consists in first thoroughly moistening or saturating a sheet which is to be coated, which I do preferably by immersion in suitable fluid, which may be cold or heated, according as one or the other condition is best adapted to produce the desired effect; then, in depositing the dampened sheet on a level table, and by suitable manipulation causing the sheet on its under side to come into contact with the upper surface of the table, by expressing air and any free fluid from between the sheet and table; and also, where found necessary, removing superfluous fluid from the upper surface of the sheet, by application thereunto of bibulous paper, or other suitable absorber or remover; and finally, in applying to the upper surface of the sheet, when in the condition produced by the second operation, a fluid material, or a fluid solution of the material or of the mixture, with which the said surface is to be coated. In practice the level table top should be made of substances not changeable in form on the application of fluid, and glass, slate, marble, or metal may be used, though I prefer glass.

"In applying the solution, which, upon drying, forms the desired coating of the sheet, and where the solution is of such a nature that but a small quantity will leave or deposit the requisite coating, I proceed in the manner of water-color artists when laying broad washes of flat tints. But when a considerable quantity or depth of fluid is required to make the desired coating, I then make use of such a frame as paper-makers term a deckle (the bottom of which may be faced with rubber), placing it on the top of the table around the paper, the edges of which the inside of the frame nearly touches. I then pour on the paper a suitable quantity of solution, which gravitates into uniform depth on the paper, it being prevented from flowing off from the paper by the deckle. In some cases the distribution of solution may be aided by the operator's use of a brush.

"In applying the solution use may be made of a reservoir, caused to traverse over the table, and delivering the solution uniformly over the breadth of the sheet while so passing; and, if desired, the solution may be made to flow into or upon a brush attached to said reservoir, and coming into contact with the sheet, while the amount of the solution delivered may be made to depend upon a suitable valvular arrangement, and the speed with which the reservoir is moved. With this reservoir may be employed the deckle, especially if considerable depth of solution is to be left upon the surface of the sheet.

"While this process was devised by me with reference to the preparation of paper for use by photographers, I do not by any means consider it as limited to such use, as sheets of material may be coated in the manner described with alcohol, aqueous, alkaline, or acid solutions, or with fused resins, oils, varnishes, or paints But care must be taken to have the fluid, with which the sheet is saturated, one with which the covering or coating solution or fluid has an affinity.

"The object of thoroughly and uniformly saturating the sheets before applying them to the table is to cause them to lie flat thereupon, so that, if the sheet is of uniform thickness, there will be no valleys in which the deposit of coating matter will be thick, or hills on which it will be thin."

It will be seen that the patent issued to Mr. Crosman is only for making tissue for use by the Swan, or other processes described hereafter. What arrangements, if any, will be made to allow the photographers of this country to use Mr. Swan's process we cannot yet say, but we are assured that no interference whatever will at present be made with any one who may feel inclined to take it up and experiment with it. A great many have already done so, and are also making their own tissue. The making of the tissue is troublesome, and it is more economical to buy it of the dealers. Great credit, honor and praise is due to Mr. Swan for his process, and the whole world is indebted to him. We had heard that Mr. Swan's and the Crosman interest had been combined, but there has been no public announcement of such a combination. We suppose it is not Mr. Swan's intention to disturb those who are working his process in this country at present. In time, he will no doubt announce his plan for licensing.

A number are now using the process, and have been doing so for two years without interference, and we believe all are safe in doing the same.

We now proceed with the details of the process as worked by Mr. Swan, adding a few foot-notes, giving our experience wherever it differs from that described so fully and so plainly by Mr. Simpson.

DETAILS OF MANIPULATION.

We shall endeavor to make plain the manipulations in the various stages of producing a carbon print.

PREPARING THE NON-SENSITIVE TISSUE.

The tissue is prepared by machinery, by which a perfect and uniform coating is secured. Each piece of paper is made into an endless band revolving round rollers, which keep it stretched, and repeatedly pass it over a surface of melted gelatine, sugar, and pigment, until a perfectly even coating of the right thickness is applied to the whole length. The trough of gelatine is kept at proper temperature by means of steam. By repeated contact with the gelatine, a thin coating being applied each time it passes over it, a much more perfect surface and even thickness of the gelatine is secured than could be obtained by any plan which applied the full thickness at once. By the arrangement adopted, waves of irregular draining are entirely avoided. These lengths of gelatine are then cut up to specific sizes, and will keep *ad infinitum*, ready for sensitizing when required.

It is important that the paper employed should possess a fine surface, and be quite free from inequalities and imperfections, in order that it may receive an even layer of the pigmented gelatine, as any imperfection in this layer may result in a blemish in the picture. It is also desirable that it (the paper) shall be sufficiently permeable by the water to facilitate its removal from the gelatine prior to development. To the quality and proportions of the color and gelatine we refer in another chapter. As the tissue is prepared and sent out ready for use by the patentee, it is not important to enter into more minute details of its preparation here.

The tissue is prepared in three distinct varieties of color; and in each scale there are three gradations of intensity, to suit negatives of various kinds. The colors are described as Indian-ink, sepia, and photographic purple.

The *Indian-ink* tissue is a pure black, nearly neutral in tone, but inclining to warmth rather than coldness.

The *sepia* tissue is of a rich deep brown, of a warm sepia tint.

The *photographic purple* tissue is of a tint resembling that common in gold-toned silver prints, generally of a purple-brown character, or in its extreme depths a purple-black.

Each of these tints is made in three qualities, to suit the different degrees of intensity in negatives, on a principle first pointed out by Mr. Swan, and which it may be well here to explain in detail.

In this method of pigment printing, although the best picture will result from the best negative, it is possible with a very intense hard negative, possessing very abrupt contrasts, to produce extremely soft and harmonious prints; whilst, on the other hand, brilliant prints may be obtained from a feeble negative possessing very little contrast or intensity. The principle upon which these effects are secured is this: The reader has seen that, as the gradations in the picture are obtained by different thicknesses of a translucent colored film resting on a white ground, the deepest shade being secured by the greatest thickness of this material, most completely covering up the white ground, it follows that the greater the proportion of color present in the film of a given thickness, the deeper will be the tint secured; and the less the amount of color present, the thicker must be the layer of the material in order to get depth. If, then, we take a sample of tissue prepared for a good negative, and print with a hard, dense negative, sufficient thickness of the colored translucent film is rendered insoluble to produce deep shadows and well-marked half-tones in the deepest gradations, long before the more delicate half-tones have been formed at all. If the printing be continued until these are secured, the lower half-tones forming the details in the shadows are obscured, sufficient thickness being rendered insoluble in the lighter of these shades to completely mask the underlying white ground. If with such a negative, however, we employ a tissue containing a much smaller proportion of color, it permits a considerable thickness to be rendered insoluble before the deeper half-tones are obscured; and in the time required for this, sufficient light has penetrated through the dense parts of the negative to render the details in the lights properly. On the other hand, by increasing the proportion of color, great contrast may be secured in the print, although little contrast may exist in the negative, as a slight thickness of the translucent material will, if it possess a large proportion of color, give great depth; and by the time the light has passed sufficiently through the thin deposit of a feeble negative to produce details in the lights, sufficient color will have been secured in the shadows to give vigor, without continuing the printing further, and so degrading the picture by rendering insoluble a further layer of color in the lights.

It will be seen, then, that by forming the picture in a thin film of insoluble matter of intense color, vigorous contrasts and perfect gradations from light to dark may be secured with a thin negative; and that by using a thicker film of insoluble matter, less intense in color, the excessive contrasts of a hard negative may be softened, thus materially ameliorating the faults of bad negatives in either direction.*

With a good negative, neither weak on the one hand, nor too intense on the other, there is no difficulty in producing perfect results, rendering between pure white and deep black every minute gradation, from the most delicate demi-tint in the lights to the least illuminated detail in the shadows.

The tissue is prepared, therefore, in each tint to suit negatives of three qualities. These are numbered, Nos. 1, 2, and 3. No. 1 possesses the smallest proportion of color, and is suited to the production of harmonious prints from negatives in which, from the nature of the subject, from under-exposure or over-intensifying, the contrasts are abrupt. No. 2 is suited to good negatives of normal character, in which the densest parts are not absolutely opaque. No. 3 possesses a large proportion of color, and is suited to thin, soft negatives, a little lacking in force and intensity. By a classification of the negatives, and the use of a suitable quality of tissue

* We say, that by the employment of a kind of tissue possessing a suitable proportion of pigments, the faults of weakness or of hardness in the negative may be *ameliorated.* We must not be understood to say *altogether* corrected; for after all that can be done in the way of compensation, defective negatives will produce defective carbon prints. And here it may be well to mention that the kind of negative which suits best for Mr. Swan's process is a negative of full average density, with full detail in the shades, such as is got by ample exposure and development. There should be *some,* although *little, absolutely bare glass;* but whatever deposit of silver there is on the deepest shades should be a pure photographic deposit, and not "fog."

for each, it will be found possible to secure more complete control over the character of the prints, and a more perfect uniformity of result than is possible in ordinary silver printing.

The tissue should be kept in a cool, dry place, packed flat, and kept under a weight. If suffered to be exposed to the atmosphere, it will be apt, in hot weather, to curl up and become unmanageably horny; whilst in damp weather it would, being a hygroscopic substance, absorb moisture.

SENSITIZING THE TISSUE.

This and other subsequent operations will of course be conducted in the "dark-room." A nearly saturated solution of bichromate of potash is employed. As the strength of a saturated solution varies with temperature, Mr. Swan prefers to make a solution of definite strength, by dissolving such a quantity of bichromate of potash as will not, during cold weather, crystallize. Such a solution is formed by dissolving one pound of bichromate of potash in twelve pounds of water.*

The tissue is immersed by drawing it (face up) under the solution (contained in a dish two or three inches deep), care being taken to avoid the formation of air-bubbles. After immersion the sheet is turned, and a flat camel's-hair pencil is employed to remove the bubbles that form on the back, which, being apt to repel the aqueous solution in small points, should be brushed over until all parts absorb properly. After the displacement of the bubbles from the back of the tissue, it is

* First wipe the surface of the tissue with a tuft of cotton, piece of cotton flannel, or soft linen, taking care not to touch the printing surface with the hands.

again turned, and is drawn repeatedly through the solution. It then has clothes-clips attached along one of the edges, and is slowly withdrawn, so that the solution drains off without being repelled from the face of the tissue, and running off in streams. If the sheet is large, a thin lath of wood may be laid along the edge of the tissue that is first withdrawn from the trough, the tissue and lath being clipped together with clothes-clips. The time of immersion may vary from one to three minutes, depending somewhat on temperature and on the facility with which the tissue absorbs the solution. As a rule, as soon as it is quite limp from the thorough permeation of the solution, it should be removed. The longer the immersion, within certain limits, the more sensitive will be the tissue; but if too much prolonged, there is danger of two serious evils. In the first place, the paper becomes rotten, the gelatine also loses toughness, and the large quantity of water absorbed renders it liable to tear with its own weight. In the next place, long immersion in a saturated solution is apt to produce a crystallized surface in drying, which, of course, renders the tissue quite useless. As a rule, perhaps, two minutes will be about the average time of immersion; but a knowledge of the degree of pliancy required will be gained from two or three experiments.

The tissue should be placed to dry in a dark room, through which a current of dry air is constantly passing. In the first stage of drying, the temperature of the air must not be above 60 or 70 degrees Fahr., for otherwise the gelatine, already softened with water, would melt. During damp weather, the air of the drying-room may be raised 10 degrees *after the tissue has become half dry*. If the drying be slow, the development of the image afterwards will be extremely slow or altogether impossible. As the result of much experience, Mr. Swan has arrived at the conclusion that keeping the sensitive tissue for a long time in a moist condition, or drying it slowly, results in a decomposition analogous to that effected by light, producing uniform and complete insolubility. We are able to confirm his conclusions from our own experiments. After *complete desiccation*, the sensitive tissue may be kept for several days. We have kept it for a fortnight without change. However, Mr. Swan strongly recommends that the tissue be used on the first or second day after sensitizing. By keeping too long, a discoloration of the print results, precisely analogous to that produced by keeping sensitive chloride of silver paper too long. The print develops tardily, and the lights are not clear. Excessively prolonged immersion in the bichromate solution of course retards drying, and should therefore be avoided. As a rule, by sensitizing in the evening, a supply of paper may be prepared for printing next day; twelve hours' suspension in a dry atmosphere being amply sufficient for the necessary drying.

Perfect desiccation, so as to make the tissue horny and unmanageable, is not desirable. It is in such a case difficult to get perfect contact in all parts in the pressure frame, and difficult to mount the tissue before development. Should the tissue by accident be rendered too dry, it is desirable to hang it for a few minutes in a damp place, when it will quickly become sufficiently pliant to permit easy manipulation. On the other hand, it is obvious that the tissue must not be too damp, or retain the slightest capacity for adhesion, or ruin to the negative would be the necessary consequence.*

* Never sensitize more than a dozen 4-4 sheets of tissue in thirty ounces of solution, and the same proportion for a larger or smaller

EXPOSURE UNDER THE NEGATIVE.

As the prepared side of the tissue is placed in contact with the negative, it is manifest, as we have just seen, that if it retained the slightest adhesiveness of surface, it would be dangerous to bring them together. Care must always be taken, therefore, not to use damp tissue.

For the exposure it is not necessary to use pressure frames with hinged backs, as the print is not, of course, examined in progress, the sole guide as to time being afforded by the photometer. The pressure of the back should be comparatively light, and the backing should be smooth and level. Fine cloth forms an excellent backing. Where the padding of the back is coarse, a piece of smooth cardboard may be placed at the back of the tissue. Too heavy pressure causes a kind of mottle of dark patches at points which have been pressed into absolute contact with the negative. If the tissue be quite dry, there can be no objection to sun printing; but if the slightest moisture were left in the gelatinous film, prolonged exposure to a hot sun with a dense negative would soften the film, and cause it to adhere. As, however, this tissue is much more sensitive than albumenized paper, printing in diffused light will generally

quantity. Then throw the solution away. Immediately before sensitizing, said solution must be thoroughly shaken or stirred; otherwise, the lower portion will be stronger than the top, and no good results can be obtained. Never put more than two sheets of tissue in the sensitizing solution at the same time, unless you can keep them *widely* separated from each other. If by dipping them vertically you can keep the entire surfaces of the sheets half an inch distant from each other, any desired number may be immersed and sensitized at the same instant.

be more convenient, as well as safer. As a rule, the exposure is from one-third to one-half of that usually required for albumenized paper. In direct sunlight we have found the exposures with different negatives vary from one to ten minutes; in diffused light, from ten minutes to an hour, or upwards. In using the actinometer, it is scarcely necessary to observe that it must be exposed to the same light as the prints, the progress of which it is intended to indicate.

MOUNTING AND PREPARATION FOR DEVELOPMENT OF THE IMAGE.

The term development is used for convenience, although it is essentially different from the operation usually known as development, in which the reduction of a metallic salt on which light has acted produces an image, the developer completing or developing an operation which light has commenced. Here light has completed the chemical action; and the operation which follows is a mechanical one, which, by the removal of the sensitive compound where light has not acted, at once makes visible the image, and prevents the further action of light.

As we have seen that the washing away of the superfluous compound must be effected at the side opposite to that which was in contact with the negative, before we can commence development, the tissue must be mounted on another piece of paper with a material which is not affected by water, in order that the paper on which the compound has rested up to the present time may be removed, so as to expose the hitherto protected surface to the water.

As the paper upon which the tissue has to be supported, during the future operations, is placed in contact with the surface which will eventually be the surface of the finished print, it is desirable that it should be smooth and free from blemish; and it should be sufficiently tough to bear the treatment necessary in hot water. Fine Saxe paper answers well.

A solution of India-rubber is used for mounting the tissue. Pure India-rubber should be cut up into fine shreds, and dissolved in pure benzole at the rate of about ten grains to one ounce of the solvent. When properly prepared, it forms a thin varnish, but it leaves a palpable film of India-rubber on the paper to which it is applied. We have at times met with samples which dissolve very tardily in benzole. Where it is found desirable to hasten the complete solution, covering the shreds of India-rubber with a little chloroform will quickly reduce them to a pasty mass, which will be readily dissolved by the addition of benzole.

The India-rubber solution is poured into a flat dish, and the paper floated (till saturated) upon it, or rather drawn over it, so as to secure an even coating on the whole surface. The paper is then hung up by the aid of clothes-clips to dry. The tissue is now floated* over the surface of the India-rubber solution in the same manner, care being taken not to allow it to sink below the surface; otherwise the back of the tissue would be coated with the India-rubber, and so retard subsequent operations. The tissue is then hung up to dry for about an hour. When the India-rubber on the paper and on the tissue is dry, the extreme edge of the tissue is

* We have always practised brushing on the India-rubber solution, *i.e.*, "Hydro-carbon Varnish," with a soft camel's-hair blender. To do this, the print is fastened to a glass by means of clips or sticking paper, and then an even coating brushed over it. It is very difficult to float the tissue.

cut off with scissors, and the two coated surfaces are carefully brought into contact and pressed together, when they cohere very tenaciously. In order to secure perfect contact and cohesion, they are next submitted to heavy rolling pressure. It is necessary to remember that any want of cohesion will issue in blisters, which will mar in greater or less degree the effect of the finished picture. It is important, therefore, that this operation be performed with care. The coated surfaces should be preserved alike from dust and from contact with fingers, or anything which could impair the cohesion of the India-rubber surfaces. In bringing the tissue into contact with the India-rubber coated paper, the tissue should be bent back, so that contact is first made with the middle of the print; the ends of the tissue being then allowed to fall after first contact. Shifting the position after the tissue has touched the paper is inadmissible; it must, therefore, be laid on straight. After being placed, the back of the tissue may be lightly rubbed with the hand or a pad, the rubbing being from the centre outwards. It is an advantage to prepare a stock of paper in advance, and to use it about an inch larger than the tissue, and to fold this upon itself, so as to form a double thickness half an inch wide. Several prints may be attached to one piece of paper.

The press used for this and subsequent operations, in Messrs. Mawson & Swan's establishment, is a powerful copper-plate press (by which a pressure of several tons can be applied), having a plate of polished steel on the bed of the press, whilst a piece of thick elastic felt is placed between the print and the roller, compensating for possible inequalities in the tissue, paper, &c., and securing perfect contact in every part of the paper and print. It is probable that the more inexpensive rolling machines might be applied to this purpose without disadvantage;

but very heavy pressure is indispensable. We find any ordinary bed-plate press, with the plate moving between the rollers, to be applicable to this stage of the process. The plate should be entirely level. In rolling, the India-rubber coated paper is laid on the steel plate, and a blanket of thick felt is laid over the tissue, which is uppermost, whilst it passes through the press.

We may remark here, that whilst the prepared surface of the sensitive tissue must be always carefully shielded from light, when once that has been covered up by mounting, it may be submitted to a dull, diffused light with impunity, care being taken that the back of the original tissue, which has now been rendered very non-actinic by the yellow color of the bichromate with which it is saturated, be uppermost. This permits the rolling of the mounted tissue to be effected in a moderately light room. The back of each print should be examined, and any India-rubber solution removed by rubbing with a piece of India-rubber. The object of this precaution is to secure uniform development. If some portions were rendered waterproof by patches of India-rubber at the back, such spots would be protected from the action of the water for a time, and would be incompletely developed when the other portions of the print were finished; the result of which would be a patch of a deeper tint than the remainder of the picture.

DEVELOPMENT AND WASHING.

The print is now ready for development. To effect this, a plentiful supply of warm water is necessary. In the Newcastle establishment, a series of three large wood-troughs are used. These are provided with hot and cold water-taps and waste-pipe. Into these troughs the prints are passed in succession. This ready supply of water and facility of securing any temperature is very desirable, as a matter of convenience, wherever the operations are conducted on a large scale. But the same result could be easily obtained on a more limited scale in photographic dishes, and having at hand a large vessel of hot water, as well as the ordinary cold water supply.

Where hot and cold water are not convenient, the following arrangement will be found very useful:

A wooden frame-work is made of the desired height, similar to a table with an open top, and a cross-piece in the centre. A and B are metal pans four to six inches deep, placed in the open top, suspended there by their rims, and heated by a gas-burner or the stove C. One of these is used for cold or tepid, and the other for hot water. It is convenient to have a thermometer in each pan.

The prints are first immersed in cold water, all air-bubbles being carefully removed. Here they are left for half an hour or more, as may be convenient, to permit the water to penetrate and soften the gelatine; after this, they are placed one by one in water of from 80 to 100 degrees Fahrenheit. This immediately loosens the backing paper upon which the tissue-compound was originally coated, and which, having now completed its office of supporting the tissue until it is no longer needed, is stripped off. It is separated from the tissue at one edge, and lifted gently away. If it still adhere tenaciously, a little longer soaking in the warm water will be necessary to effect the removal of the paper; but this is always a bad sign. The back surface of the tissue, opposite to that which was exposed, is now uncovered; and the next operation is to remove all gelatine, pigment, and chromic salt which have not been rendered insoluble.

The operation of developing, up to this period, has been conducted in a subdued or yellow light. As the sensitive surface is now exposed, it is obvious that strong white light should be avoided until the bichromate has been washed out of the film. This is rapidly done. A large portion has been removed whilst the print was soaking; and now that the gelatinous compound is exposed to the warm water, the salt is rapidly diffused in the water. The process of clearing may be accelerated by allowing a gentle stream of the warm water to fall on the surface of the print, or by laving the water on to it with the hands, so as to produce slight attrition between the surface and the water. This, however, is not necessary, as, if the print be left face down in the warm water, it will be found, in the course of from five minutes to a quarter of an hour, to have parted with nearly all the superfluous gelatine and color, presenting the image in all its proper gradations, and only requiring a little further washing to complete the operation.

The usual temperature for development is from 80 to 100 degrees, Fahrenheit; but there are circumstances which modify this. If, from over-exposure, the picture appear too dark, or from some tendency to insolubility in the compound, the image appear slowly, the temperature may be raised, when necessary, even to 150 degrees Fahrenheit; but high temperature must not be used until all the development has been effected that can be effected by water of a lower temperature.

The development is best *commenced* at as low a temperature as possible; and then, as soon as the image is fully made out, the print should be removed to cold water, in which the residue of bichromate will be washed away without risk of injury to the delicate half-tones, which would, with an under-exposed print, disappear in hot water. After two or three hours' immersion in cold water, the prints are one by one re-immersed in water at 80 or 90 degrees. Those which show signs of under-exposure are very carefully rinsed in merely tepid water, say 80 degrees, to clear away the soluble gelatine and adherent color; after which they are suspended to dry. The more fully-exposed prints remain longer in the warm water, in fact, until they become light enough. Any that are over-exposed are put into hotter water, and are allowed to remain until the depth is sufficiently reduced. By judicious management of the development, using merely tepid water (not over 80 degrees), at the commencement of the operation, any under-exposed prints are discovered and saved. Then, by the use of hotter water to the more fully-exposed prints, these are speedily lightened to the required degree, and thus very few prints are lost either from under- or over-exposure.

When sufficient gelatine and coloring matter have been removed, and the prints are fully developed, they are hung up to dry. In the developing operation, several prints may be placed in one vessel; but as the image, although no longer soluble in water, is still slightly gelatinous, it is liable to abrasion; and care should be taken to avoid the prints dragging over each other, or over the bottom of the dish.

There are a few precautions to be carefully observed in development. It is most important to preserve uniformity of action. If, for instance, an air-bubble form, at any period before development is complete, the film of air protects the spot from the solvent action of the water, and the picture will be darker in that place. If the picture be suffered to float with the face partially out of the water, the same thing will happen. It is desirable, therefore, to keep the face downwards until the operation is completed, and to remove air-bubbles whenever they form. It should further be remembered, in observing the depth of the picture, that it is now seen on a ground considerably degraded by the coating of India-rubber, which gives the paper a brown tint, and that when transferred to pure white paper, it will possess much greater brilliancy.

TRANSFERRING THE PRINTS.

The picture, up to the present time, presents an image in which right and left are reversed. It is now necessary, therefore, to transfer it from the paper which has supported it temporarily for the purposes of manipulation, to its final resting-place, in which operation the right and left will resume their proper relations. The image may be either transferred to a sheet of cardboard, so as to require no further mounting, or to paper; in the latter case, it is simply in the position of an ordinary print, and will require subsequent mounting. Each method has certain specific advantages, but generally the transfer to paper is to be preferred.

Transferring to Cardboard.—The face of the dried print

is very evenly coated by floating, or by means of a flat camel's-hair brush, with the following preparation:

Gelatine, ...	2 ounces.
Glycerine, ...	½ ounce.
Water, ...	1 pint.

The gelatine should be melted and carefully cleared of air by long heating, and skimming the froth; after which the glycerine is added. It will at all times, of course, require melting by heat and straining through wet flannel or muslin before use; it is then applied evenly to the surface, either by floating (which is best), or with a broad camel's-hair brush, and afterwards hung up to dry. Sometimes in coating the print with the gelatine solution, there is a disposition manifested by it not to adhere on all parts of the surface of the print, but to "creep" off in certain spots. When this happens, there can be no adhesion at those points between the gelatinous coating and the pictures; consequently, the film, composing the picture, will be torn off *from such places*, when the rubbered Saxe paper is removed. To avoid this non-adhesion and "creeping," it is best to coat the prints in a warm room, where the hot gelatine will not chill, nor "creep," upon being brought in contact with the surface of the print, as it would do if the print were in a cold apartment. When dry, the print is trimmed to the required shape. A piece of stout cardboard of the required size, pure in color and fine in surface, is passed through clean water, and then drained. Upon the moistened surface the print is laid, face downwards, exactly in the position it is designed to occupy, and the card is removed to the rolling-press and placed on the polished steel plate, print-side downwards, the side on which the print is placed being in contact with the plate, and a felt blanket on the back of the card; it is now submitted to a heavy rolling pressure, and then put aside to dry.

The quality of the cardboard and the exact condition of dampness are of considerable importance. It must be perfectly moistened all over, as, if any point or patch were omitted, the adhesion of the print in that place would not be secured. There should be an absolute film of water on the surface, so that as each part is submitted to the rolling pressure, a wave, infinitely small, however, is driven before the pressure, effectually displacing air, and securing perfect contact. It is, however, undesirable to have excess of water. There should be no delay in applying the pressure after the print has been placed in contact with the moistened surface, inasmuch as the gelatinous, although insoluble image, by absorbing moisture and becoming soft, might, under the heavy pressure, lose something in sharpness.

As each print is passed through the rolling-press, it is placed upon the last, and when the pile is completed, a weight is placed upon the whole heap. By adopting this course, the prints dry without warping or cockling; and at the expiration of about twenty-four hours the print is ready for the final operation.

This consists in removing the paper which has supported the image during the operations of developing and washing. The picture must be quite dry before the operation is attempted. A piece of clean cotton-wool is saturated with pure benzole, and the caoutchouc-coated paper which covers the print is rubbed pretty hard with it. An edge of the caoutchouc-coated paper is then gently raised with the point of a blunt knife, care being taken to commence at a black part of the picture where the film of the compound forming the image is thickest. The raised edge is then taken hold of, and pulled so as to tear it gently and steadily off the print. Instead of removing the paper with an upward or lifting motion, it is better to turn it backwards, so that the strain is in a horizontal direction, as there is, in this method, less danger to the surface of the print at any point in which the adhesion in mounting is imperfect. As a general rule, especially when the benzole is used sparingly, the paper brings away with it all the India-rubber coating; but any traces remaining may be rubbed away with the finger or with a piece of India-rubber. It is best always to rub a sponge, dampened with benzole, over the surface of the picture as soon as the Saxe paper is removed, even when there are no perceptible adhering spots of varnish on the print. Under ordinary circumstances, the picture is now finished. If required for coloring, the print may be coated with plain collodion, or a suitable sizing preparation.

It is important to remember that defective manipulation in the mounting operation seriously mars the beauty of the finished picture. It is necessary that the pressure should be perfectly uniform in order to secure evenness of texture in the surface of the picture. If the transferring coating of gelatine were laid on in uneven patches or streaks, the effect of this will be apparent in patches or streaks of greater brightness or dulness of surface, the thickest parts receiving the highest pressure, and consequently having the brightest surface. Any unevenness of pressure in rolling will produce a similar result.

Transferring to Paper.—The manipulations here are very similar to those which we have just described, but are a little more easy. It is not necessary to trim the print to its proper size or shape, as this will be done in the final mounting. The mounting papers are carefully immersed in water, air-bubbles being brushed away, and then laid one upon another while in the water; they are then drawn out in a pack, and are suspended to drain for some hours, or submitted to pressure to remove the superfluous water; a perfectly even film of moisture is thus secured. The transfer is effected by laying the print face up on the steel plate of the press, and over the print is laid the moistened paper, and on that a felt blanket. The press is then "pulled." The print is next immersed for an hour in a bath, containing five per cent. of alum, and is afterwards well washed in water and dried, after which it is uncovered as when mounted on cardboard.

By transferring to paper, it will be observed that facility is afforded for performing the last-mentioned operation, by which an additional source of stability is secured. The only possible source of deterioration in the prints produced by the method we have described, exists in the thin coating of gelatine with which the print is attached to its final support. By means of moisture and friction the print could be removed; this, it is true, is destruction, not fading or instability in its usual sense. But it is happily possible to remove even this susceptibility to injury. Although the transference of the print direct to cardboard has the advantage of making an exceedingly neat finish to the mounting (the print being slightly recessed in the cardboard), and although it has the further advantage of reducing the number of operations required to complete the picture, yet Mr. Swan greatly prefers, and almost invariably adopts, the method of transfer to paper, chiefly because this method secures the most uniform adhesion, and because it allows the gelatine (used to cause the print to adhere to the paper), to be rendered water-proof — a property not possessed by the prints mounted direct to card. One of the means used by Mr. Swan to render the gelatine insoluble is quite novel, and constitutes one of the first applications of his discovery of the property possessed by salts of the sesquioxide of chromium of rendering gelatine insoluble. A solution of common alum has, to a certain extent, the power of waterproofing the prints, and generally fixture with alum is quite sufficient. Where, however, more thorough waterproofing is demanded, the prints, after transfer, should be treated with a one per cent. solution of chrome alum. Mr. Swan has shown us some prints very successfully transferred without a press. The transfer was effected with the gelatine solution ordinarily used, to which has been added one-twentieth of a ten per cent. solution of chrome alum. Prints intended for coloring in water colors should be chrome-fixed.

Mr. Swan generally adds a small proportion of a white pigment to the gelatine with which the transfer is effected, in order to give brilliancy to the whites of the picture, and to avoid the intervention of a transparent film between the under surface of the print and the paper to which it is attached.

THE SENSITIVE COLLODIO-GELATINE TISSUE.

The method of carbon printing with this tissue is better suited to the amateur than to the professional photographer. It involves more trouble than the use of the paper tissue, but the results are very beautiful; and as the photographer, in employing it, mixes his own preparations, he has certain points in color and intensity more completely under his own control than he could have in purchasing a ready-prepared tissue.

To prepare the Sensitive Collodio-Gelatine Tissue.—Take a sheet of plate-glass, free from blemishes or scratches, and clean it perfectly, finally rubbing the surface with a saturated solution of beeswax in ether. This is then wiped off with a clean cloth, leaving a scarcely perceptible coating of the wax. This coating may be omitted; but it tends to facilitate the future removal of the tissue from the glass.

Now coat the glass with a plain collodion, giving a thick, tough, transparent film. The pyroxyline should be of the kind which yields a film free from opacity or opalescence. About ten grains in an ounce of solvent, consisting of equal parts of ether and alcohol, will answer the purpose. This film is, of course, suffered to dry before applying the tissue compound.

Next make a solution of gelatine and sugar, as follows:

Pure gelatine, ...	2 ounces,
White sugar, ...	½ ounce,
Water, ...	8 ounces.

The kind of pigment to be employed, and the proportion in which it is to be added, will depend much on circumstances, into the details of which we enter in another chapter; but it is especially important in the preparation of this kind of tissue, that the pigment employed be so finely divided that no subsidence will take place during the period the tissue compound remains in the fluid state upon the glass. The preparation in this state may be kept ready for use. It should be kept in a well-corked, wide-mouthed bottle; in hot weather it is apt to decompose if kept long. It may, if desired, be poured into a flat dish to the depth of about half an inch, and, when nearly dry, cut into shreds, and thoroughly dried; in which state it may be kept without risk of injury. When required for use after drying, it must be soaked again in eight parts of water.

The proportion of gelatine and of sugar given is that which is found to answer best under ordinary circumstances. But these proportions will be influenced by the quality of the gelatine, the temperature, and varying conditions, in regard to which experience must be the guide. In very dry weather, for instance, the proportion of sugar may be increased, its chief office being to give pliancy and elasticity to the tissue, and prevent the horniness of the gelatine when perfectly desiccated.

To prepare the tissue compound for use, heat must be applied until it is quite fluid, when one part of a saturated solution of bichromate of ammonia must be added to every ten parts of the gelatinous compound, after which the whole should be strained through flannel. It is desirable, after the chromic salt has been added to the gelatine, to avoid applying a greater heat than is necessary to preserve fluidity, as excess of heat tends to produce spontaneous insolubility. About 100° Fahrenheit will generally answer the purpose. It should be further remembered that frequent and continued application of heat to gelatine destroys its setting powers, which would render the preparation useless.

The thickness of the tissue, and the proportion of the mixture necessary in forming it, depend very much on circumstances. If the tissue be too thin, the finished picture will not possess its proper depth of shade in its darkest parts, unless it has had an unusually large proportion of coloring matter. If too thick, drying is retarded, and it is intractable in mounting and other manipulations; it will also require a longer time in development. As different qualities of gelatine will produce different results, something must be left to experience in determining the amount of the tissue-compound necessary to form a given amount of tissue; as a general rule, however, it may be stated that about two ounces will be required for each superficial foot.

Immediately previous to the preparation of a sheet of tissue, the piece of "patent plate" glass should be placed in a levelling stand, in a perfectly horizontal position, a spirit-level being used in the adjustment. The tissue-compound, warmed to 100°, should be strained through a piece of moist flannel or muslin, and when the compound is ready the plate should be warmed until it is of the same temperature as the compound. The proper amount is then poured on the collodionized plate, and caused to flow over its surface, a glass rod being used to spread the solution. A little care is necessary to prevent the formation of bubbles of air, which, when once formed, are not easily disengaged from the viscous solution, and, unless eliminated, result in defects in the tissue, producing white spots in the picture. The coated plate is then left on the levelling stand until it is quite set. As will easily be seen, a very little inclination will cause the coating to run into uneven waves, or to accumulate and form a greater thickness in parts, the disadvantage of which is manifest.

When once thoroughly set, the plates may be placed away in an upright position to dry. The more quickly the drying is effected, provided heat be not applied, the better. The temperature should not exceed 60° or 70° Fahrenheit, as a higher temperature may cause the gelatine to run, and form uneven waves. A very low temperature, which would retard drying, is, of course, undesirable. A damp place is especially to be avoided, as the protracted drying*, caused by a damp atmosphere, materially tends to the production of a spontaneous decomposition and general insolubility of the tissue. In a dry, well-ventilated dark-room, kept at a temperature of about 60° Fahrenheit, drying will generally take place within twelve hours, and without any danger to the solubility of the tissue. It may be found desirable in damp weather to use a drying box, containing chloride of calcium, sulphuric acid, or other substance having great affinity for water.

When the tissue is dry, it is ready for printing; it is, therefore, removed from the glass and placed in the pressure-frame, with the collodion surface in contact with the negative. The proper exposure is ascertained by the aid of the photometer described on another page. Before development, the tissue is coated with India-rubber solution in the same manner as the paper tissue referred to in another chapter, and is, in like manner, mounted on paper coated with India-rubber. It is then developed, washed, dried, and transferred as already described; the film of collodion, in this instance, forming the surface of the finished print.

Instead of coating the glass-plate with collodion, it may be rubbed with ox-gall, or with the solution of wax before mentioned, and coated with the sensitive tissue compound. When this is dry, it may be coated with collodion, removed from the glass, and treated in the manner already described. Or it may, instead of being coated with collodion, have a sheet of wet paper applied to it, and pressed in contact so as to adhere. It is then suffered to dry, and treated as the paper tissue in all respects, its only difference consisting in the fine surface communicated by the plate-glass, which becomes, finally, the surface of the transferred picture, and possesses a little more delicacy of effect than that produced by the ordinary paper tissue.

CARBON PRINTING IN THE SOLAR CAMERA.

By Antonio Montagna.

The process I use to make enlarged carbon prints consists of six successive operations:

1st. *Choice and Preparation of the Glass.*—The glass

must be pure white, and the surfaces flat and parallel. After cleaning it in the ordinary manner, I cover it with alcohol 30 grammes,* water 8 grammes, and nitric acid 10 drops. I dry it, and throw on the surface a small quantity of dry soap powder, the excess of which I remove by gentle touches of a fine brush.

2d. *Coating with Collodion.*—My collodion consists of ether 150 grammes, alcohol 80 grammes, and gun cotton 6 grammes, and the glass is covered with it in the usual manner. After drying, I proceed to

3d. *Application of the Sensitized Gelatine.*—This preparation consists of

Distilled water, ...	90 grammes.
Pure gelatine, ...	10 "
Bichromate of ammonia, ...	1.25 "
Liquid Indian-ink, ...	15. "

To this is added a sufficient quantity of some aniline color; for instance, Magenta, to give a warm tone to the picture.

This preparation is poured over the dry collodion film, and dried in a horizontal position, so that the film attains the same thickness throughout.

4th. *Exposure in the Solar Camera.*—The printing is done from behind. The side not coated is exposed to the image of the negative, and the time of this exposure is only half as long as that required to obtain a print on chloride of silver paper.

5th. *Development.*—This is simply done by repeated washings in warm water, and, when the soluble gelatine is removed, washing is continued with cold water; then the glass is dried in an oblique position.

* The gramme is very near 15 grains, Troy.

6th. *Mounting of the Print and Separation from the Glass.*—As soon as dry the print may be retouched, if necessary. This retouching must be done by looking at it by transmitted light. To mount it I paste, by means of gelatine, one or more sheets of white on very light tinted papers or cardboard on the print, taking great care to remove all air-bubbles which may be entrapped between the paper and the moist glue. I place it under pressure, and leave it for twenty-four hours; then I cut the edges loose with a penknife, and leave it exposed to the air, when in a short time the paper spontaneously detaches from the glass, the print adhering firmly to it.

When the paper used has a proper tint in harmony with the subject, and with the carbon and Magenta color of the print, it has all the appearance of a varnished oil painting.

The author liberally offers any more information on the subject to all who desire it.

NOTE.—Notwithstanding this process is a little laborious, and more adapted for amateur performance than for the regular profession, it contains so many practical hints that it was thought its publication might be of benefit to those who desire large carbon prints.

SWAN'S ACTINOMETER.

Before proceeding to manipulatory details, it is desirable to mention the mode of meeting another serious difficulty which has hitherto stood in the way of carbon printing. As the film of bichromated gelatine and carbon is *black*, there is no darkening during exposure to indicate the progress of printing. No perceptible change is made in the film by the action of light.

In the uncertainty arising from this circumstance it has been necessary to guess the exposure, or estimate it by a consideration of the strength of the light and the density of the negative. If carbon printing were to be carried out on an extended scale, it was manifest that some more accurate mode of proceeding was necessary. Mr. Swan and Dr. Vogel have met the want by the instruments described below.

Swan's actinometer consists of a small box, in which is inclosed a piece of sensitive paper, carefully screened from the action of all light except that to which the operator submits it. This box is provided with a sliding lid, in one aperture of which is fixed a small screen of glass, which has been collodionized, excited, exposed, developed, &c., so as to form a miniature negative, nearly opaque at one end, and nearly transparent at the other. Under a small section of this (of an appropriate degree of translucency), the sensitive paper is exposed to light. Another portion of the lid consists of yellow glass, underneath which the sensitive paper can be pushed, and examined without danger of injury from the light; the slightest tint of the portion upon which light has acted being readily distinguishable through the yellow glass from the white portions upon which light has not acted even. It also possesses an arrangement for bringing under the screen for exposure successive portions of the sensitive paper, as each colored portion has done its office.

The actinometer, in its most perfect form, is provided with the graduated screen, or a series of screens, each of different density, corresponding to the density of various negatives. But it may be very easily worked with one screen, in which case the screen is very translucent, and is termed a unit screen; with this, several repetitions of a constant tint—and that almost the first remove from absolute white—are given in each printing—two, three, four, or more, according to the density of the negative.

These, however, are points of detail in which each operator will adopt the method which suits him best. It is assumed that the negatives will be sorted and classified; and, where it is necessary, the printing qualities of each, and the section of screen, or number of unit tints, required on the actinometer will be ascertained by one or two preliminary trials, and marked on the negative. With a simple system of classification and registration it will be easy to secure sufficiently uniform results. In some large printing establishments a similar system is pursued in silver printing. The negatives are classified, and the whole of one class being exposed at one time, it is only necessary to examine the progress of printing under one negative, which becomes practically an actinometer; and when it is completed, it is known that all the others, possessing like qualities, and having been submitted to similar conditions, are completed at the same time.

The sensitive paper for the actinometer may be prepared by almost any formula, provided uniformity be observed. Plain Saxe paper, immersed for ten minutes in a ten-grain solution of chloride of sodium, may be kept ready for use. This, when required, may be floated for two minutes on a forty-grain solution of nitrate of silver, and will be found to answer every purpose.

The apparatus is supplied with the necessary mechanical appliances for ready change and examination of the sensitive paper, and is found perfectly practical, being at once easy to use, and efficient for the purpose for which it is designed.

VOGEL'S NEW PHOTOMETER.

The photometer or actinometer used by Mr. Swan, and described above, was found to be wanting some of the advantages required, but which are entirely secured by Dr. H. Vogel's admirable invention described below.

The objects of this ingenious device is to enable the operator to judge of the proper time to expose a negative in the carbon printing process, and to determine the exposure when making negatives in the glass-room. Those who have experimented in carbon printing know the difficulty of determining the proper exposure, too short or too long a time destroying the print altogether. The use of this little instrument overcomes that. In making negatives on days when the light is variable, difficulty is often experienced in securing the proper time of exposure. This is also the case when making copies, views of interiors, and landscapes. A proper understanding and use of Dr. Vogel's photometer will be found of immense advantage in such cases. The length of exposure of any desired picture, in any desired weather, can at once be ascertained. Mr. Swan has adopted this photometer.

We shall now proceed to describe it.

It consists of a box A A, provided with a lid B B, as shown in the drawing. The lid consists of a frame, *a*,

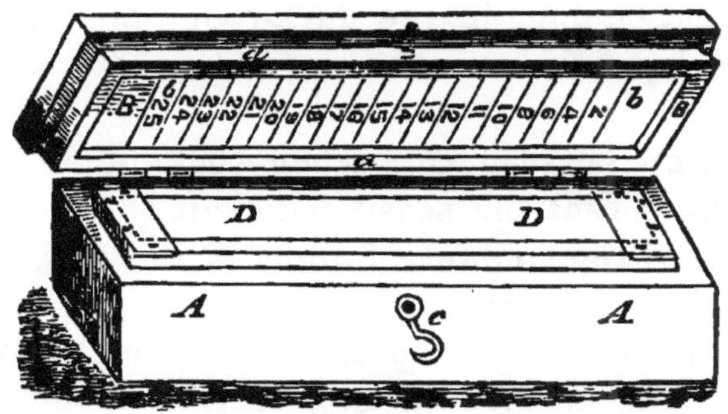

by which a glass plate, *b*, is held; on the upper side of this glass plate is secured a series of thin strips of paper, which are arranged mathematically in layers; each lower layer projecting, like steps, beyond the layer above. This step system thus produced, represents a semi-transparent medium, the transparency of which decreases by degrees towards the thicker end. Black figures upon the under side of the lower strip indicate the number of layers arranged above each such figure. The whole cover, with the paper system in it, can be folded down and fastened by the small hook, *c*, and is provided with a second cover of wood, to protect the glass from injury, when not in use, and to make and close the exposure.

Within the box is a sliding false bottom, D, which is, by means of a steel spring, pressed upwards against the scale described above, when the lid is closed. Upon this false bottom a number of strips of paper are placed, which have been sensitized by immersion in a saturated solution of bichromate of potash of 1 oz. bichromate to 30 oz. water.

In order to get these in place, the bottom of the box is opened, the spring removed, the false bottom taken out (and may be used as a guide to cut the strips of sensitized paper), the strips placed in the box, the false bottom dropped in upon them, the bottom closed, and the photometer is ready for use.

This must be done in the dark-room, and the fingers should be dry. Strips sensitized in this way will keep a month.

It will now be seen that when the apparatus is exposed to light that the sensitized strip changes color in proportion to the amount of light it receives, the most light passing through the spot marked 2, the next 4, the next 6, and so on, and the change will be rapid or slow, according to the intensity of the chemical action of the light. The black figures admit no light through them, and after exposure appear on the sensitized strip as light figures on a dark ground. A strip can only be used once, and after use is easily removed and the one under it exposed ready for use.

In the carbon printing process the instrument is exposed to the light with the negative which is to be printed from, and when it shows six degrees, the first quarter of the negative is covered with black paper between the negative and the carbon tissue, or otherwise; when eight, the second; when ten, the third; and when twelve, the fourth. In this manner the single parts to 6, 8, 10, and 12, have been printed. The print is then developed, and notice taken of which part shows the best intensity, *i. e.*, 6, 8, 10, or 12, and ever after, the time of exposure for that negative in the same light is established. When examining the scale it should be held to the light, and the eyes should be shielded from bright light with the hand or otherwise. No. 2 will appear first, and the others with decreasing distinctness. When examining the scale hold it about eight inches from the lamp, allowing a bright light to shine perpendicularly upon the paper, and examine the latter obliquely, keeping the eyes from the flame in the direction of the figure 25.

The scale is also provided with hands and letters (omitted in our drawing), and to which attention should also be given, as they facilitate observation. Dr. Vogel does not take the highest figure visible as the copying grade, but the next one below it. For Rowell's tissue he finds for a medium negative that 11 or 12 is the proper degree, or for a dense negative 14 or 15; for Swan's, 15; and that it is better to take a degree more than less, *i. e.*, better to over-expose than to do the reverse. The Photometer is equally useful in the photo-lithographic, photo-engraving, enamel, and aniline processes.

For determining the time necessary for the exposure

of the negative in the camera by the collodion process, Dr. Vogel gives the following directions:

While preparation is being made to take the picture, expose a plate to the light at the place where the model is to sit, one minute, timing it with the watch. Now expose the photometer for one minute in the same light, and note the degree indicated on the sensitized slip.

We believe this instrument will be found useful to every photographer, and, as we understand, it will be furnished at a low price; no doubt all who desire to progress and improve will possess themselves of one or more. Keep the glass clean above the scale. Don't touch the scale with damp fingers.

THE CHROMIC SALTS.

Either chromic acid alone, or various of its salts, may be used to render certain soluble organic bodies insoluble after exposure to the action of light. Practically, for a variety of reasons, the bichromate of potash, or the bichromate of ammonia, is found to answer best. Bichromate of potash, as being the cheapest, is more generally employed, otherwise bichromate of ammonia has certain advantages. It is a little more sensitive, and has been said to yield a tissue less liable to spontaneous insolubility. The latter quality, however, seems doubtful, and certainly requires verifying; as, for more than one reason, it seems probable that it will, on the contrary, produce a tissue more prone to spontaneous decomposition. In the first place, this tendency will necessarily accompany extreme sensitiveness. Mr. Swan finds that dampness in the sensitive tissue is a chief cause of the spontaneous change which produces insolubility, and as bichromate of ammonia is more greedy of water than bichromate of potash, the tissue will more readily absorb moisture from the atmosphere, and thus secure the condition favorable to spontaneous change. The difference in the solubility of the two salts is the chief element of advantage in favor of the latter. Bichromate of potash is soluble in about ten times its weight of water, at 60 degrees; bichromate of ammonia is soluble in about four times its weight of water, at the same temperature.

The double chromate of potash and ammonia has been recommended as possessing certain advantages over the bichromates. M. Emile Kopp, who first suggested its use, prefers it as more convenient. Mr. Carey Lea points out that its especial advantage is found in the fact that it is less liable to spontaneous decomposition, and not seriously less sensitive to the action of light. It is not necessary to prepare the crystallized double salt, but simply to neutralize a saturated solution of bichromate of potash with liquid ammonia. The advantage of the bichromate in point of sensitiveness is however very great. It will be found advantageous, therefore, wherever the greatest sensitiveness is required, to use a bichromate in preference to a neutral chromate.

The salts of uranium have, in some instances, been proposed in place of those of chromium, but the reactions do not correspond, and practically this substitution is inadmissible.

PHYSIOLOGICAL EFFECTS OF CHROMIC SALTS.

It is important to those using chromic salts to have some knowledge of their effects on the health if used carelessly. The bichromates taken internally are active poisons, but Dr. Alfred Taylor remarks in his work on poisons, that notwithstanding their extensive use in the arts, well-observed instances of poisoning by means of a bichromate are rare. Dr. Cloet, who made a careful investigation into the condition of health of the people engaged in a bichromate manufactory, states that, taken internally, it is not poisonous in such minute doses as cyanide of potassium, about fifteen grains being necessary to cause death in a healthy adult person; but it is in coming in contact with the mucous membrane, or with a slight abrasion of the skin, that its most injurious action is found,—obstinate and dangerous ulcerations, issuing occasionally in complete destruction of the part, ensuing. It is observed that when used with care, no danger whatever need be apprehended, as it is quite innocuous on the skin when there is no abrasion; no absorption of the poison taking place, except by the mucous membrane, or through a wound of some kind. The important point for photographers to observe is, to avoid contact with the bichromate where there is any scratch or lesion of the skin, and to avoid contact with the eyes or nostrils with fingers which have recently touched the chromic salt. Dr. Cloet says:

"This salt (bichromate of potash) in small doses, say a few grains, acts as a purgative; if in larger doses, say fifteen grains, it acts as a poison. A workman in a factory put some bichromate into a barrel of cider, by way of joke. The cider was rendered dark in color, but still the other workmen drank of it, and were all affected with severe colic and diarrhœa. Disease of the nostril has been produced by workmen who, having stained their fingers with the salt, have put them into the nostril.

"In transforming neutral chromate of potassa into bichromate, by means of acid, the vapor arising carries with it an infinity of pulverulent molecules of the product, which spread through the workshop. This cloud of particles is easily visible in a ray of sunlight. The molecules inspired give a bitter and very disagreeable taste to the palate; but as profuse salivation is the result, the chromate is thrown off in the saliva, and has not time to inflict any permanent injury. If, however, respiration be made through the nose, the molecules are dissolved in the layer of secretion which lies on the membrane, creating a violent pricking, suffusion of tears, and irresistible sneezing. In time, the membrane begins to be thrown off, and portions of it are carried into the handkerchief used in blowing the nose; this process, when once started, goes on so rapidly, that after a period of six or eight days the septum becomes thin, permeated with openings, and is ultimately detached altogether. Snuff-takers escape this evil.

"On the skin, in its normal state, and intact, the bichromate exercises no baleful influence; the hand may, in fact, be plunged into a hot concentrated solution of the salt, without fear; the hand may also be covered with the salt for an entire day without any observed effect; but if the skin is torn or abraded, however triflingly, by the prick of a pin for example, a sharp pain is felt on contact of the salt, and if it be left in contact with the wound, the caustic character of the salt is brought out intensely, the cutaneous tissue is decomposed, and violent inflammation is established. These symptoms are accompanied with intense pain, especially in winter, when the cold is severe; the action of the salt does not cease until the cauterization has penetrated to the bone.

"When the skin is abraded, and the bichromate has produced ulceration, the best treatment is to wash the part thoroughly with a feebly alkaline water; then, if inflammatory action follows, to poultice, and afterwards freely apply subacetate of lead in solution."

Dr. Taylor recommends for cases where the poison has been taken internally, emetics and carbonate of magnesia or chalk, mixed with water into the consistency of cream.

THE PIGMENTS EMPLOYED.

Considerable latitude in the choice of pigments is permissible, as almost all those employed by the painter are available in preparing the tissue for printing by this process. Where especial effects, resembling artist's drawings, are required, which, in reproductions will often be valuable, it is quite possible to produce them. The effect of a drawing in lead-pencil may be imitated by using graphite as the pigment; red chalk may be imitated by Venetian red; for sepia and bistre effects these pigments themselves may be used.

For most purposes, however, a fine black, either neutral, or inclining in tone to brown or purple, will be preferred. Fine lampblack, or good Indian-ink, will, in such case, generally form the basis of the coloring matter. If the color required be a pure neutral black, the addition of a blue pigment is necessary to neutralize the brown tint of Indian-ink; and, where necessary, coldness is corrected by the addition of some warm color. The selection of this color will be governed by the tint desired, and by the considerations of permanency. Many of the most beautiful tints are most fugitive. Carmine, for instance, is unstable; and some samples are injured by the action of the chromic salt. Crimson lake is a valuable color, but it is not strictly permanent. Indian red is a very powerful and very permanent color. Venetian red is also permanent. For blue, ultra-marine is quite satisfactory as regards permanence.

In judging of colors for this purpose, it should be borne in mind, that the actual effect of color employed is chiefly seen in middle tint. It is difficult to distinguish much difference between a blue-black, a brown-black, a purple-black, a rosy-black, &c., in the extreme darks of a picture; but the tone is easily distinguished in middle tint, and, as, a rule, warm half-tones are the most pleasing. It should also be remembered that a weak picture will often look brilliant in a warm tone, whilst a vigorous print will look feeble in a cold color.

We have stated before, that by the addition of a large proportion of color to the gelatine, a vigorous print may be obtained from a feeble negative, and by the use of a small proportion of color, a hard and intense negative may be made to yield soft prints. As a normal proportion, however, for good negatives, two per cent. of carbon is sufficient. Of course, the proportion of pigment required is different with different pigments, and depends upon the opacity and colorific power of the color employed. Mr. Swan prefers the use of insoluble pigments, as when the tissue is prepared with soluble colors the prints are apt to lose a little of depth and force, if they are subjected to prolonged washing.

THE GELATINE.

Gelatine, as found in commerce, is a very variable substance, and is often impure. It is difficult to give a rule for its selection; but, speaking generally, the gelatine sold for culinary purposes answers well for carbon printing. Common glue is not suitable, and the best samples of gelatine used in cookery are unnecessarily expensive. Different gelatines vary considerably, both in the proportions required, and in the results they produce. Some samples of commercial gelatine, and some of glue, show a tendency to dissolve in cold water. These are unsuitable for the process. Impurities, such as alum or acid, are highly objectionable.

A HINT ON THE PREPARATION OF SOLUTION OF INDIA-RUBBER IN BENZOLE.

A scientific friend has directed our attention to a curious circumstance which he has observed in preparing a solution of India-rubber in benzole; and, as many of our readers have met with difficulties in the direction which we shall now mention, we throw out the hint here. Our friend had great difficulty in preparing a solution of the gum even moderately clear, and, as he wished to have some perfectly bright for particular work, he made many experiments upon the subject. On thinking over the matter, he came to the conclusion that it is the moisture of the India-rubber which communicates the milky appearance to the benzole solution. On boiling the liquid for a minute or less, all this moisture was found to be given off in addition to a little benzole, and, on cooling, the liquid remained perfectly clear, although rather thickened. The addition of a very small quantity of water caused the resumption of the original opalescent appearance, thus proving clearly the cause of the peculiar properties of the solution.

On repeating this experiment before us, we suggested to our friend that the addition of any very hygroscopic substance which would remove the water from the liquid should produce the same result as boiling, and in a very much more simple way. A few fragments of fused chloride of calcium were, therefore, added to some of the opalescent liquid, and with the result of rendering the liquid perfectly clear after a few minutes' standing upon the salt. The chloride of calcium can then be removed, as it is quite insoluble in the benzole, and the liquid then preserved for use.

There is another fact in connection with this subject which it may be interesting to mention. When a dark-colored sample of India-rubber is used in preparing the benzole solution, the brown color may be almost completely removed by the addition of a little water to the liquid. An emulsion is then formed of a milky appearance, and which leaves, on evaporation, a film but slightly colored.—*Br. Jour.*

COLORING CARBON PRINTS.

Carbon photographs, finished as described in the chapter on Mr. Swan's process, admit of coloring in oil, water, or powder colors, with the greatest facility, and without risk of damage; the manipulation is easier than that upon albumenized silver prints.

POWDER COLORS adhere very readily to the surface of these prints. By breathing on the picture, a still more adherent surface is obtained. If greater depth of tint than can be secured in one application of the colors be required, a coating of a very thin varnish may be applied after the first tinting, and on a second application of the color, considerable depth and brilliancy will be obtained.

WATER COLORS.—If, after the final transfer of the print, the gelatine employed in the operation were not rendered insoluble by a solution of alum, or the chrome salt, the use of water colors on the print would be attended with danger, as the moisture would be absorbed by the gelatine used in transferring, and this film which forms the back of the picture would be apt to be disturbed by the abrasion of the pencil. When the print is properly finished, however, according to the instructions we have already given, there is no such danger. The water colors take kindly without any preparation, washing well, and permitting tint to be worked over tint without difficulty. The surface may be rendered still more pleasant for working on by the application of any good "sizing preparation." Nothing can be better for water colors than the carbon print so treated. The plain carbon print so treated acquires an even, clear surface, losing all gloss without any loss of depth or transparency, which is very pleasing.

Oil Colors.—The best mode of preparing a carbon print for the reception of oil colors consists in sizing it with isinglass. A solution of about two per cent. of isinglass in equal parts of hot water and spirits of wine, carefully applied (not too hot) to the surface of the carbon print, with a flat camel's-hair brush, yields a surface upon which oil colors work admirably.

Retouching Carbon Prints.—In the ordinary process of retouching carbon prints, to remove small imperfections, it is only necessary to use the proper color in the usual way; but if a little gelatine, with a trace of a chromic salt, be employed with the color, or if any good "Sizing Preparation" be employed as the medium, instead of water, the color will, when dry, become insoluble, like the rest of the picture. If the retouching be effected with the same materials before transferring the print, it will, when the picture is finished, be under the image, and no inequality of surface, usually apparent after touching, will be seen. This method permits the character of a print to be considerably modified, without the manipulation being obtrusively apparent in the finished picture.

PRACTICAL NOTES ON THE CARBON PROCESS.

TRANSFERRING WITHOUT A PRESS—TRANSFERRING WITHOUT GELATINE—CARBON PRINTS ON PORCELAIN GLASS—CARBON NEGATIVES.

Another and very interesting method of transferring prints has been tried practically, after an idea of Mr. Swan, by Dr. Vogel. The characteristic of this method is, that a press can be dispensed with.

The picture, resting on caoutchouc, is immersed in a warm solution of gelatine.

Gelatine, ...	6 to 8 parts.
Glycerine, ...	2 to 2½ "
Water, ...	100 parts.

As soon as all the air-bubbles have been removed, a piece of fine paper is immersed, and both the paper and the picture are removed from the dish by drawing them over the corner of the same; both are suspended and left to dry; they can then be easily trimmed, pressed on moist pasteboard, and separated by means of benzine.

Dr. Vogel has recently made the curious discovery also that carbon prints can be transferred without the use of gelatine. This process is much more simple than Swan's. The time which is gained by doing away with gelatinizing, drying, &c., &c., cannot be too highly estimated.

The operation is carried out in the Royal Academy at Berlin, in the following manner:

Common paper, as white and smooth as possible, is dipped for two minutes in cold water; it is then dried a little between blotting-paper, and the developed picture, after having been dried, is laid on it, picture side down, and smoothed over with the hand. After this, it is placed in the press, the moist sheet downwards, and on it a piece of felt. It is drawn once through the press, and suspended for drying. The rollers must work very evenly, or the pictures are apt to become wrinkled.

For small pictures, a copying press will suffice. The moist paper is placed upon blotting-paper, the picture is placed on top of this, it is pressed down a little with the hand, and then pressed in the press for about two minutes.

After drying for thirty minutes, the picture is dipped for one minute into a solution of chromate of alum, 1:300. After this, it is dried again. The time required for drying at 16° Beaume, is about one hour. The separation is performed as described above. The benzine is applied on the side of the caoutchouc paper.

The conditions for the success of this process are: a soft caoutchouc paper, a good caoutchouc solution, well-sized paper, and strong pressure.

Careful drying is important, and strong rubbing in with benzine. If, however, in separating, some parts of the picture should tear, and the whole picture should be difficult to separate, then it is better to stop the separation at once, to place the picture into a glass or tin dish, to place a piece of plate-glass upon the picture in order to press it, and to pour benzine upon them until they are covered.

To avoid the evaporation of the benzine, place the dish with the pictures into a larger dish, which fill for about one-quarter of an inch with water, and now place over the dish containing the pictures, another one inverted, which will dip with its sides in the water, and prevent evaporation. The pictures remain in this for about ten minutes, when they can be easily separated.

All pictures must, on account of the caoutchouc which attaches itself to them, be rubbed off with a piece of flannel saturated with benzine.

When the picture on caoutchouc has been thickly gelatinized (12 per cent. gelatine), it will be easy to remove it from the paper as a pure film.

This circumstance led me to experiment, to transfer the carbon (pigment) picture to glass, and the experiment succeeded perfectly. For this purpose, I covered the picture resting on the caoutchouc, thickly with a solution of gelatine:

Gelatine,	12 parts.
Glycerine,	4 "
Water,	100 "

And I glued it in this manner to a previously-warmed plate of glass, avoiding, of course, all air-bubbles. After having left it for drying, I removed the paper with benzine, and the picture remained perfect on the glass.

It is advantageous to cover the paper on the back, after drying, with a solution of—

Chromate of Alum, ...	1 part.
Water, ...	300 parts.

Dissolve off, as usual, with benzine. Should there be any difficulty in removing the picture, it is advisable to place it in benzine in the manner described above.

The transfer to porcelain (opal glass), is performed in the same manner. This gives very pretty effects, but great care must be exercised in dissolving them off.

It is self-evident, that this process is of great importance for the email and porcelain photographs.

If, to the pigment of the first gelatine sheet, an email color has been mixed, a picture will be obtained which can be burnt in.

Another interesting circumstance, I only wish to indicate. In the pictures on glass, we evidently have a carbon (pigment) positive. From this, it will be easy, by repeating the process, to produce a carbon negative.

We would thus be enabled to multiply our negatives, and to produce, instead of the perishable silver negatives, permanent ones by the carbon or pigment process.

FAILURES, FAULTS, AND REMEDIES.

Before proceeding further, a brief recapitulation of the causes of failure in Mr. Swan's process, and the remedies, may not be out of place.

Spontaneous Insolubility of the Tissue.—This, as has been said, arises chiefly from *slow drying*, or, *long keeping in a damp place.* The addition of substances to give elasticity, such as glycerine, which retard the drying of the excited gelatine film, also tend to produce spontaneous insolubility. Heat, in conjunction with the moisture, increases the tendency. The use of too much bichromate of potash, or too prolonged immersion in the solution of bichromate, will produce spontaneous insolubility. Immersion in very hot water, prior to development, is at times conducive to insolubility, also drying the tissue in an impure atmosphere, and especially one vitiated by the burning of gas. Some samples of gelatine are said to become readily insoluble; but this requires confirmation.

Tardy Solution of the Superfluous Gelatine in Development.—The same causes which will produce spontaneous insolubility, when present in less degree, cause tardy solution of the unaltered gelatine, and slow development. The more rapidly the tissue has dried, and the more horny and perfectly desiccated it appears, the more readily, as a general rule, the superfluous gelatine and pigment are removed by warm water, and complete development is effected. When the development is slow, hotter water may be employed; but care should be taken that the free soluble bichromate has first been removed by tepid water.

Bichromate of Potash Crystallizing on the Tissue in Drying.—If the tissue be suffered to remain too long in a saturated solution of bichromate of potash, the salt will crystallize on the surface during drying, and the tissue will be useless. The remedy, of course, is the employment of a weaker solution, or a shorter immersion in the full-strength solution.

Uneven Development.—If the print be suffered to float to the surface of the warm water, allowing portions to become dry; or if some portions of the paper forming the original basis of the gelatine, be suffered to become detached long in advance of the remainder, so that the warm water acts directly on the soluble matter in patches, the result will be uneven development, the portions last uncovered remaining darker than the rest of the print; and it will be difficult to equalize the tint, even by long-continued development.

Blisters During Development.—If, in mounting the tissue with the India-rubber solution, perfect contact in all parts be not secured, blisters will arise in the course of development, which will show as marks or defects in the finished print. They are also caused by small holes in the paper, or air which remained between the two varnished surfaces, or by want of contact in the rolling. If from the first cause, they will dry down, disappear, and do no harm. If from the second or third causes they may be pricked with a fine, sharp needle, from the back of the paper, and so rendered harmless. There is, however, once in a great while, a very refractory blister which will spoil a print, or which can only be removed by scraping, and retouching with Indian-ink, after the print is mounted. With great care in placing the two varnished surfaces together, before rolling, and care in rolling with a pretty heavy, steady pressure, the blisters may be entirely avoided.

Over-Exposure.—An over-exposed print will develop tardily, and continue, under ordinary treatment, too dark. After all the soluble chromic salts are removed, the temperature of the water may be raised, and by long soaking in hot water the depth may be reduced considerably. Mr. Swan has found that immersion for a short time in a very weak solution of chloride of lime, or of hypochlorite of soda, or in chlorine water, or peroxide of hydrogen, rapidly reduces a print, by decomposing a portion of the insoluble chromo-gelatine compound, and restoring it to its original condition of solubility. The action is, however, too energetic to be of much practical use in the reduction of over-printed pictures. Protracted immersion in hot water is the best remedy.

UNDER-EXPOSURE.—An under-exposed print develops rapidly, the lighter half-tones rapidly disappearing. When this tendency is seen, quickly removing the print to cold water will arrest the progress of development, and by skilful manipulation and attention, and the after use of almost cold water (say under 80°), a brilliant print may be secured.

WEAK AND FLAT PRINTS.—When a feeble print is obtained from a good negative, it may arise either from the use of a tissue containing too small a proportion of color, or from the tissue being old and partially decomposed by slow drying. If the negative be weak, the use of a tissue containing a large proportion of color will yield a vigorous image. Increased vigor may be obtained from an ordinary sample of tissue, by sensitizing it on the paper side of the tissue only, instead of immersing the whole. Printing in direct sunshine aids in obtaining a vigorous print.

HARDNESS AND EXCESSIVE CONTRAST.—This may arise from the use of an unsuitable negative, or from the injudicious use of too hot water on a lightly exposed print, or from the use of tissue containing an excessive proportion of color, especially in conjunction with under-exposure. Sensitizing the tissue on the prepared side will tend to produce softness, even with a dense negative.*

* Mr. Swan observes that in printing from negatives somewhat deficient in softness of gradation—in which case there is a tendency

An Uneven Texture in the Finished Print, some portions looking more glazed than the rest.—This defect arises from unequal and insufficient pressure in transferring. This unequal pressure may arise from the coating of India-rubber being uneven, or, more probably, from the coating of clear gelatine being applied in uneven streaks, or from uneven texture of blanket, or uneven pressure.

Portions of the Image tearing off in Transferring.—This will arise from the face of the print being *imperfectly coated with gelatine*, or from the paper or board to which the print is transferred having an imperfectly moistened surface, or from not being dry when the paper is removed, or soiled by fingering or dust.

A Green Tint Pervading the Blacks is caused by imperfect washing of the print, by which traces of soluble chromic salt are left in the image.

Unequal Sensitiveness.—This arises from the tissue having imbibed the bichromate solution unequally. If, in immersing the tissue, one portion remains dry while the rest is wet, that portion will be least sensitive, and will form a light patch in the picture. If the tissue is raised out of the bichromate in such a manner that streams of the solution run down the sheet, there will be in the print patches or streaks of a darker color, corresponding to the streams of the solution. The attachment of a strip of paper along the lower edge of the tissue, immediately after it has been hung up to dry, helps

to abruptness in the transition from white to the lightest tint,—it is advantageous to expose the sheet for a moment to diffuse light, so as to produce a uniform tinting of the slightest degree possible. In vignetting, Mr. Swan regards this as almost indispensable. It may also be well to remark here, that the weakest tissue (No. 1) should be employed in vignetting, and that the vignetting screen should be very softly graduated in tint.

to drain off the excess of solution, and tends to equalize the sensitiveness.

The Gelatinous Coating Runs in Sensitizing.—This will happen if the bichromate solution is too warm, and the tissue kept too long immersed. During summer it is necessary to keep the bichromate solution as cool as possible, and to sensitize in the coolest place that can be procured.

Dark Spots.—If a piece of tissue is printed under too heavy a pressure, dark spots or patches appear in the half-tones. This is most apt to occur if the tissue is limp, and the pressure of the back of the printing-frame not only strong, but uneven from coarse padding.

A Sparkling Appearance in the Print after final Transfer.—This arises from the transfer process being imperfectly performed, the paper being either too wet, or too slight pressure used, or the blanket not sufficiently yielding to diffuse the pressure equally over all the surface of the print.

Cracking of the Film.—This will occur sometimes after the print is entirely finished, generally only during cold weather and is caused by sudden changes of temperature from hot to cold, and *vice versa*, while the developed print is drying. Avoid this.

MANY MITES FROM MANY MINDS.

Be very *certain* that both of the varnished surfaces are perfectly dry before putting them together.

The sooner the tissue paper can be separated from the Saxe paper, in the tepid water bath, the better it will be for the picture; and, for this reason, never take into the tepid water bath more than a dozen pictures at one time, for fear they might soak there *too long* before they could be separated, and so, possibly, give trouble.

The "hydrocarbon varnish" (or India-rubber solution), and the "transferring solution" (or benzine), are very volatile and inflammable; should be kept tightly corked, to prevent waste from evaporation; and must never be used near a naked flame or fire.

The varnish brush, to be kept soft and straight, should be suspended from a hook on the underside of the lid, in a tin brush cup, which should always contain enough of the "transferring solution" to keep the brush saturated.

Pictures technically called "vignettes," of the size of life, or cartes-de-visite, may be made on the tissue without difficulty.

The utmost care and attention must be used not to allow the tissue to be struck by light, after it is sensitized, either before placing it under the negative or in any of the subsequent manipulations. Remember that the sensitized tissue is *much* more sensitive to the action of light than any silvered paper, and that any want of care, in this particular, will certainly be punished by the entire failure of the whole operation. Several, who have used the tissue, have lost many pictures by neglecting this often-repeated caution.

As many prefer ascertaining the density of the sensitizing solution by the use of a hydrometer, instead of weighing the crystallized salt, and as the hydrometers in common use, unless of high cost, are not made with much accuracy or uniformity, it is recommended that each photographer who prefers to use the hydrometer, should carefully make one sensitizing solution, as herein directed, from the crystallized salt, and ascertain its density *by the particular hydrometer he has in use.* By this means, he may be sure of the strength of his sensitizing solution without the trouble of frequently weighing the crystals of his bichromate salt.

No definite time can be given for immersing the tissue in the sensitizing solution, because the time should vary with the season of the year and temperature. In warm weather, in summer, *one minute*, or even less, is sufficient; in a moderate temperature, in spring and fall, say a little less than *two minutes;* and in cold weather, in winter, *three minutes.* The only disadvantage from an unnecessary long time in the solution is, that it loads the tissue with a superfluous quantity of water, which requires too long a time to dry out or evaporate, and makes the tissue very tender. But the tissue cannot be "*over-sensitized*" (in the sense in which the term is used in sensitizing paper for silver printing), by too long immersion in the solution.

In making the second transfer, Mr. Swan recommends a solution composed of *two ounces* of gelatine, *half an ounce* of glycerine, and *one pint* of water. As it contains no sugar, it may, under some circumstances, be found preferable to the other formula.

Another sensitizing solution has been recommended, and is made by dissolving *three ounces* of bichromate of potash in *thirty-five ounces* of water, and adding strong ammonia until the solution becomes alkaline. Use litmus paper to test its neutrality, and when neutral add *a quarter of an ounce* of ammonia, to be sure that the solution is strongly, but not too strongly, alkaline.

It is recommended to all, to try it. It is thought, by some persons, to possess advantages over the other in rendering the tissue more flexible, and causing it to lay more closely to the negative. On the other hand, these qualities are disputed.

It has been found that good card-board sometimes answers a *better* purpose than the felt-cloth, in making both the transfers. It is, therefore, suggested to all to try it; but, where card-board is substituted for the felt-cloth, it is obvious that the press and rollers must be true, and work with great accuracy.

In making "vignettes," twenty-five or thirty grains of chloride of barium may be added to every ounce of gelatine employed in the gelatine solution, which, on coming in contact with the tannin solution, will be converted into sulphate of baryta, and will give more clearness to the white background.

It is a good plan, though not very important, to filter the sensitizing solution.

The solution of gelatine should be filtered, while hot, through fine linen cloth.

It is best to make the first transfer, and develop the prints, soon after lighting, otherwise, they may *sometimes* become partially or wholly insoluble. As a general rule, the tissue may be lighted in the morning and the pictures developed in the evening; but, *occasionally*, from some unknown cause,—it has been known to occur but once in two years—the tissue, after having been lighted under a negative, will become insoluble if kept too long before being transferred and developed.

The last transfer may be made by pulling apart with the fingers, without the application of "transferring solution" or benzine, if the prints are allowed to dry twenty-four hours after the last pressure.

The clothes-pins used should always be clean.

The tissue is better dried quickly, but artificial heat must not be used to hasten it. Hang it, if possible, in a draught of air or well-ventilated room.

It is a good plan to take a sponge or soft brush saturated with water, and remove any adhering bichromate solution from the back of the print while it is drying, and take care not to touch the blackened surface. This will prevent the appearance of insoluble spots in developing. Thoroughly shake the sensitizing solution before using.

Don't put too many sheets in at a time.

If you make a stock solution of bichromate, keep it well corked to prevent evaporation.

Sometimes, in coating the print with the gelatine solution, there is a disposition manifested by it not to adhere on all parts of the surface of the print, but to "creep" off in certain spots. When this happens, there can be no adhesion at those points, between the gelatinous coating and the pictures; consequently, the film, composing the picture, will be torn off *from such places*, when the rubbered Saxe paper is removed. To avoid this non-adhesion and "creeping," it is best to coat the prints in a warm room, where the hot gelatine will not chill nor "creep," upon being brought in contact with the surface of the print, as it would do if the print were in a cold apartment.

In preparing the moist paper to place the print upon for the final transfer, first soak it thoroughly in water, and then blot off or press off all that is free.

There is no difficulty whatever, in avoiding blisters in developing the prints; all that is required is, first, that the caoutchouc paper and solution for coating the tissue be prepared with highly volatile solvents (such as rectified benzine, and probably the *very lightest* of the spirit from petroleum, will answer equally well); any of the less volatile hydrocarbons remaining mixed in small proportions with the lighter through imperfect rectification, are mischievous. Second, the proportion of caoutchouc in the solution and on the paper, should not be too small. Third, heavy pressure must be applied after the tissue is laid down on the caoutchouc paper: this is most important; that is to say, if the tissue and paper are laid together dry, if I may use the term. Of course, if the caoutchouc solution, or the tissue, or paper, is fluid when they are brought into contact, air can be excluded, and perfect cohesion obtained without pressure, but then a long time must be allowed for evaporation of the caoutchouc solvent. But probably the dry method will be preferred, and with that method strong pressure is indispensable, in order to avoid blisters. Fourth, in developing the prints, the paper on which they rest must be carefully handled, so as to avoid creases and rumples. Wherever a crease or incipient break in the paper occurs, there will be a blister. Fifth, avoid the use of very hot water. To that end, use tissue easily soluble; that is to say, tissue which is fresh, and that has been so quickly dried as to retain the normal solubility of the gelatine almost unimpaired.

When removed from the bichromate solution, if the back of the tissue be drawn over a glass rod the superfluous solution will be removed, and prevent insoluble spots in the print.

Never be in too great haste. Follow the given rules, and you will succeed. Cleanliness and care are quite as essential in this as in the silver process.

Remember that bichromate of potash is quite as poisonous as cyanide, and, getting it into cuts and abrasions of the skin, should be avoided.

Bubbles and blisters, in the tissue, arise from the presence of air in the pores, which suddenly expands when the tissue is suddenly placed in hot water, and swells into bubbles under the film of India-rubber. If the tissue is immersed for a few minutes, first in cold water, previous to development, face up, the air is expelled by the water in minute globules, and blisters avoided.

Do not be in too great haste to separate the picture.

The immersion in water also removes the chromate from the paper, which thus becomes insensitive to light, and permits the development to be made in the daylight.

It is not found best to use a brush on the surface of the newly-developed print. It endangers the whites. Moving it dexterously back and forth in the water, face up, will remove all the superfluous carbon.

Better to gradually increase the temperature of the water during development, than to have it too hot at first, and have to decrease it. Do not, however, increase the temperature too rapidly, or uneven development and blisters will occur.

The addition of half a drachm of balsam of fir to the pint of "hydrocarbon varnish," shaking it repeatedly during the day, will secure greater adhesive qualities.

The felt cloth used should be thick, of close texture, and soft.

Mr. W. J. Kuhns announces, that by simply washing the print with aniline colors before transferring, almost any tone or color may be given the print, and with beautiful results.

Liesegang recommends protecting the prints with a thin film of plain collodion.

HISTORICAL NOTES ON CARBON PRINTING.

Photographic Carbon Printing may be said to have commenced with the labors of M. Nicephore de Niepce, in 1814. His pictures were produced by the action of the solar rays upon certain hydro-carbons, which were rendered insoluble in the usual menstrua, wherever they had been submitted to the influence of light.

In 1839, Mr. Mungo Ponton announced, in the *Edinburgh New Philosophical Journal,* a process of producing images by the action of light on paper which had been impregnated with a solution of bichromate of potash.

M. Becquerel shortly afterwards investigated the action of chromic salts on organic substances under the influence of light, and arrived at the conclusion that the coloration and insolubility were due to the reaction which took place between the chromic acid and the sizing matter in the paper, as, on using unsized paper, the effect could only be produced in course of time. Upon this, he used another modification of the process: employing a paper sized with starch, he subsequently treated the image, obtained by the aid of the bichromate, in a weak alcoholic solution of iodine, and so obtained a blue tint. In 1852, Mr. Fox Talbot patented a method of photo-engraving, in which he availed himself of the reaction between organic matter and chromic acid under the influence of light.

In 1855, M. Poitevin announced the first carbon process. He proposed to render the reaction between a chromic salt and organic matter available in producing photographs in permanent pigments, the image being formed either of coloring matter miscible in water, or of a fatty ink.

In the same year, M. Lafon de Camarsac announced his discovery* of methods of producing photographs in enamel colors, in which, although very vaguely stated, are involved the elements of a carbon process.

In 1857, Dr. Phipson conceived that certain volatile oils, such as those contained in the oil known as *huile de Dippel*, or those which accompany naphthaline, and which blacken when exposed to the light and to the air, might be used to obtain photographs.

In September, 1857,* Mr. Thomas Sutton proposed adding the finest pulverized charcoal to an albuminous solution of bichromate, so as to form a black mixture of about the consistency of common shoe-blacking, which was then applied to paper. This was to be exposed under a negative, and the unaltered material subsequently removed by repeated washings, and a carbon print produced.

Later in 1857, M. Testud de Beauregard, a gentleman who had devoted much attention to the production of photographs without salts of silver, patented a process, not differing from that of M. Poitevin in principle, but having some modifications in detail. In his specification he re-states the fact as already recognized, that if a mixture of gum or gelatine, bichromate of potash or ammonia, and an insoluble coloring matter, such as carbon, black-lead, vermilion, indigo, &c., be exposed to light, it is rendered insoluble, the coloring matter or pigment being imprisoned or retained by the mixture, and that if such exposure be effected under a negative, the portions not acted upon by light may be washed away, leaving an image in pigments.

Mr. Pouncy appears to have been the first in England

* Comptes Rendus, June 11, 1855.
† "Photographic Notes," vol. ii.

to produce carbon prints by means of photography.* The earliest examples exhibited were shown at the April meeting of the London Photographic Society, in 1858: but the author declined to describe the process. An allusion to it and its results, without details, had appeared in one of the journals a few weeks earlier, viz:

"He prepared the paper, or other surface, for having the picture produced on it, by applying over its whole surface the coloring matter which formed the picture, and together with this coloring matter applied a substance which is acted on by the light. The following is the manner in which he proceeded when printing positive pictures on paper from negative pictures: He coated the paper, or surface which received the picture, with a composition of vegetable carbon, gum arabic, and bichromate of potash, and on to this prepared surface placed the negative picture, and exposed it to the light in the usual way. Afterwards, the surface was washed with water, which dissolved the composition at the parts on which the light had not acted, but failed to affect those parts of the surface on which the light had acted. Consequently, on those parts of the surface the coloring matter remained in the state in which it was applied, having experienced no chemical change. Sometimes, for the vegetable carbon, he substituted bitumen."

MM. Henri Garnier and Alphonse Salmon, in August, 1858, laid before the French Photographic Society, a "Method of Carbon Printing," in which paper was treated with a strong solution of citrate of iron, and after drying in the dark, was exposed under a transparent positive. The parts acted upon by light were rendered insoluble, whilst the parts protected by the dense portions of the cliché remained soluble and hygroscopic; and

* It is right to place on record here that this has been denied. Mr. Portbury, in a letter to the *Photographic News*, Nov. 23, 1860, and personally at a meeting of the Photographic Society, Nov. 4, 1862, claimed the production of the first carbon prints, stating that he was at the time an apprentice with Mr. Pouncy.

when tested with powdered lampblack, plumbago, or other pigment in fine powder, and then breathed upon, the color adhered to the hygroscopic portions, forming the shadows, but not to the lights, which had been rendered insoluble by the action of light. After washing in clean water, the print was perfected.

At the same meeting, M. Gabriel de Rumine described a method of obtaining prints analogous in some respects to that of MM. Garnier and Salmon, but rather resembling the patented method of M. Testud de Beauregard. He treated paper with a solution of gelatine and bichromate of potash, and, when dry, covered the surface with black-lead. After exposure under a negative, the print was removed to a dish of boiling water, which removed the soluble portions of the gelatine and the adhering color. M. Brebisson, a few months later, proposed a similar method, applying the color after instead of before exposure.

Mr. Charles Seeley, M.A., editor of the *American Journal of Photography*, proposed a method with gum, carbon, and a bichromate, mixed, and applied to paper, which he found successful. On learning that M. Poitevin had anticipated this method by several years, he ceased to prosecute it.

In the following year, the Duc de Luynes' prize was awarded, an event interesting in the history of carbon printing. In 1856, this nobleman had offered certain prizes for improvements in connection with photography, including one of 2000 francs ($400), for a method of producing permanent prints. A number of processes were forwarded to the Commission, consisting of modifications of silver printing processes; but these were put aside as unsatisfactory, and attention chiefly given to the methods of carbon printing, the competition being practically confined to the processes of M. Testud de Beauregard, MM. Garnier and Salmon, and Mr. Pouncy.

The decision of the Commission was, that the process of MM. Garnier and Salmon and that of Mr. Pouncy were about equal in result, the latter requiring, however, an exposure of nearly four times as long as the former; but being at the same time simpler in manipulation. M. Poitevin had not entered the competition or contributed specimens; but recognizing in him the originator of all the processes of carbon printing, the Commission felt bound to acknowledge his merit. Instead, therefore, of awarding the whole amount as one prize to any individual, it was resolved to divide it, awarding a gold medal, value 600 francs, to M. Poitevin as the originator of carbon printing; a similar medal to MM. Davanne and Girard for contributions to the improvement and stability of silver prints; a silver medal, value 400 francs, to MM. Garnier and Salmon, for their carbon printing process; and a similar medal to Mr. Pouncy, for his process.

A further prize of 2000 francs, offered by the Duc de Luynes for the same purpose, was awarded to M. Poitevin in 1862. In 1867 he received the prize of 8000 francs, offered for the best mechanical printing process with fatty ink, based upon photography.

Early in 1864, a suggestion for the production of carbon prints was made by Mr. Obernetter, of Munich. In his process, paper is treated with a solution of sesquichloride of iron, chloride of copper, hydrochloric acid, and water. After drying, the paper is exposed under a negative, and then developed in a solution of sulphocyanide of potassium, sulphuric acid, a little of the sensitizing mixture and water, and then washed. The image is formed of sulphocyanide of copper: if the print be exposed to an atmosphere of chlorine, the image is converted into a chloride of copper. The prints so produced were of a chestnut brown.

We now come to another definitely-marked feature in the history of carbon printing. In all the efforts hitherto made, there was one signal defect,—the absence of perfect gradation of tone from light to dark. A certain granular grayness in patches represented half-tones in some of the processes; but the tendency was to abrupt steps from white to black, any definite approach to delicately-marked gradation being rarely obtained. This was believed to be owing to the nature of the materials employed: the notion prevailed that the finest mechanical subdivison of a pigment could not equal in delicacy the fineness of the deposit obtained by the reduction or precipitation of a metallic salt. We know now that this was not the cause of the difficulty, which was due to the mode rather than to the material, as will be seen from the explanation which follows. When the surface of a film of bichromated gelatine is exposed to light, all portions upon which light acts in the slightest degree, whether through half-tones or shadows, are rendered insoluble at that surface, the only difference being that the light penetrates deeper in the shadows, and therefore produces a thicker layer of insoluble matter. When the exposed print is placed in a solvent, the whites, having been protected from the action of light, are laid bare at once; and the water then penetrating laterally, dissolves the soluble layer underneath the thin insoluble film, which forms the half-tones; these being thus deprived of their contact with the paper, float away, leaving only the deep shadows, in which the light has penetrated quite through the film. The picture thus consists of masses of black and white, without true gradation. The only wonder is that any approach to half-tone at all was ever secured by such a mode of operating. Such gradation as was obtained appears to have been dependent upon the fact that a thin solution and an absorbent paper were employed, the bulk of the sensitive material being in absolute contact with the fibre of the paper; so that any portion, on which light had acted at all, was not readily removed from the paper. Certain it is, that the principle of washing away the unaltered material from the side opposite to that which was exposed to light is vital to the perfection of gradation in carbon printing. As the discovery of this principle has been the subject of a little misapprehension, it is interesting to trace it here to its first enunciation, and mark how its gradual recognition, and the discovery of practical means of applying it, have led to the perfection of carbon printing.

The first recognition which we find of this principle is in a paper on the use of linseed oil as a sensitive agent in photographic engraving, by M. L'Abbé Laborde, communicated to the French Photographic Society in July, 1858.* In this paper, the Abbé announced his discovery that linseed oil which had been treated with litharge was sensitive to the action of light, becoming insoluble under its influence, in like manner to asphaltum. In the course of his experiments with this substance, he had discovered that the insolubility caused by the action of light commenced at the surface exposed, and gradually penetrated through the film in direct proportion to the intensity of the light. He acknowledged the loss of half-tone, although he did not suggest a mode of meeting the difficulty.

In November, 1858, Mr. J. C. Burnett, in a communication on carbon printing, pointed out the same fact still more clearly, and indicated the direction in which effort must be made in order to overcome the difficulty. After

* Bulletin de la Societé Française de Photographie, August, 1858.

speaking of the application of a mixture of gelatine, bichromate, and pigment to paper, he says:

> "It must be observed that the possibility of producing half-tones by this plan rests on the power of the insolubility—causing actinism to penetrate, with a certain degree of facility, the mixture of pigment with bichromate and gelatine, or gum, the gelatine or gum being in consequence rendered insoluble, to a greater or less depth on different parts of the picture, according to the varying depth to which the actinism has been allowed, or had time, to penetrate; this, again, being dependent on the varying translucency of the different parts of the negative."*

It was at this juncture of the history of carbon printing, we learn, that Mr. Swan commenced his experiments. Without being aware of Mr. Burnett's suggestions, and prior to the publication of Mr. Blair's letter (presently to be mentioned), he attempted to obtain half-tone by coating a plate of glass with a mixture of lamp-black, solution of gum arabic, and solution of bichromate of potash. After drying the coated plate, he exposed it in the camera, with the uncoated surface of the glass turned towards the light passing through the negative and lens. The plate was then washed in water, with the view of removing, from the back of the sensitive coating, those portions which the light had not rendered insoluble. The experiment was not successful (probably in consequence of the too feeble action of the light); it shows, however, that Mr. Swan not only, *ab initio*, recognized the true principle upon which the production of half-tone in carbon-printing depends, but, furthermore, applied it in a very elegant manner.

Early in the following year, Mr. Blair, of Perth, made the same discovery;† for it is worthy of note that in

* "Journal of the Photographic Society," vol. v, p. 84.

† "Photographic Notes," vol. iv, p. 45.

each case the recognition of the fact upon which the loss of half-tone was based, seems to have been the result of independent observation. Having applied a thick coating of the sensitive gum and carbon to his paper, he noticed, after exposure, that "the outer crust was more sunned and hardened than the inner," and that, by the time the paper was sufficiently steeped, the inner surface, which had been least sunned, was too soft, and was washed away, carrying with it the "outer crust." It occurred to him, therefore, that if he could get the inner surface rendered insoluble first, he could overcome the difficulty. He accordingly attempted to print through the coated paper, so that the inner surface was acted upon first, leaving the soluble portions on the outer surface to be removed by washing. This gave a certain amount of success; but the exposure was long, and the print looked granular from the texture of the paper through which the light passed. Subsequently, he tried the use of waxed paper as a support for sensitive material. This lessened the exposure, and in some degree lessened the granulation; but the lights, consisting of waxed paper, were not good in color, and certain practical difficulties remaining, the method did not come into general use. The principle of securing half-tone in carbon prints was now known, but an efficient and practical mode of applying it remained unknown. To the subsequent modes of utilizing this principle, we shall refer presently.

Towards the close of 1859, M. Joubert called attention to a process of carbon printing which he styled phototype. An example, published in the journal of the Photographic Society, in June, 1860, showed that the process possessed much promise of excellence, the results far surpassing anything till then seen. That they could be printed with facility in large numbers was manifest from the circumstances under which the specimen was issued. The details were not published, and remain to this day the secret of the inventor.

The next step in the history of carbon printing is based upon the discovery of another effect produced on certain bodies by the action of light, and is again due to M. Poitevin. It will be noticed that the processes already described depend on the action of light in rendering a soluble body insoluble. This is the effect on asphaltum, upon a mixture of some organic bodies and a chromic salt, and upon some other substances. In the new process discovered by M. Poitevin, a body previously insoluble is rendered soluble and hygroscopic by the action of light. The first results of this process were shown to the *French Photographic Society*, in July, 1860, the process having been patented the month previously. The details were published in the following November.* In this process a mixture of perchloride of iron and tartaric acid, ten parts of the former to four of the latter, dissolved in one hundred parts of water, is the sensitive preparation. It is poured on a plate of glass which has been previously coated with collodion or other suitable material; it is then left to dry in the dark, and becomes spontaneously insoluble. Submitted to the action of light, however, it again becomes hygroscopic. After exposure under a negative, if breathed upon, the parts upon which light has acted become moist and tacky, in degree and depth proportioned to the action of light. Finely-powdered carbon, applied with a brush, adheres to the image in greater or less proportion, just in the degree in which moisture is absorbed, thus giving an image with a just gradation of half-tone. For producing enamels, a vitreous powder has to be applied; and for

* "Photographic News," vol. iv, p. 331.

the production of images in printing ink, a fatty acid or a resin has to be applied. The print, if in carbon, was subsequently washed with water containing a little hydrochloric acid, and a piece of paper coated with gelatine applied to its surface, attached to which the collodion film, with the image upon it, was removed from the glass. In some cases the sensitive fluid was applied direct to the glass, and when the picture had been produced by the application of carbon to the exposed film, a coating of collodion was applied, and the whole eventually transferred to gelatinized paper. This was the first process in which a collodion film was used for the purpose of transferring the carbon picture from a glass plate to paper.

In the following November,* M. Fargier brought before the French Society a process for which, in the September previous, a patent was obtained. It consisted of an ingenious combination of previously published discoveries. A plate of glass was coated with a mixture of gelatine, bichromate, and carbon, and, when dry, exposed under a negative. The exposed film was then coated with tough, plain collodion; and, after allowing the film to set, the whole was plunged into warm water. This dissolved the portion of gelatine which remained soluble, and detaching the film from the glass, removed the unaltered pigment and gelatine, leaving an image attached to the collodion film with perfect gradation of half-tone from white to black. The film was next attached to a sheet of gelatinized paper, collodion side uppermost. The results were exceedingly beautiful, far surpassing in delicacy and gradation anything which had previously been produced in carbon printing.

* "Photographic News," vol. iv, p. 390.

In the spring of 1863,* M. Poitevin made another important advance in carbon printing, based upon his last discovery,—that soluble organic substances might be rendered insoluble by the action of metallic salts, and recover their solubility under the action of light. It will be seen that in this operation the reactions are just the reverse of those in the chromo-gelatine processes. Paper, coated with the sensitive salt and pigment combined, simply needs exposing to the direct action of light under a transparent positive, and washing in water. The dark coating forming the shadows retains its insolubility; and in the half-tones, the film being rendered soluble through a part of its thickness, is washed away in due proportion, whilst the light having penetrated quite through the film in the whites of the picture, the color is washed away entirely, leaving the bare paper.

The mode of proceeding is as follows: Five or six parts of gelatine are dissolved in one hundred parts of water with gentle heat, and to this the necessary proportion of carbon (or some inert pigment), is added, and the paper is coated with the mixture. When required for use, these are impregnated with a solution containing ten parts of perchloride of iron, and three parts of tartaric acid in one hundred parts of water. This paper is left to dry in the dark, when it becomes insoluble, even in boiling water. It is then exposed under a positive cliché, and, under the influence of light, becomes soluble, commencing at the surface of the film. A short exposure is sufficient, and the print is then immersed in warm water, which removes the soluble matter, leaving the print with its true gradations of light and shade. It is now necessary to remove the tint given by the iron salt to the paper, and this is done by washing in a dilute

* "Photographic News," vol. vii, p. 124.

solution of hydrochloric acid. The print is then rinsed and dried. To prevent injury to the gelatine film, which would become further soluble by the action of light, it is rendered wholly insoluble by any of the known methods, such as immersion in a solution of alum, bichloride of mercury, &c. This process has not hitherto been successfully worked.

In the same communication, M. Poitevin, mentioned another mode of carbon printing. Paper treated with perchloride of iron and tartaric acid, without pigment or gelatine, is exposed under a positive cliché. The parts treated with these salts possess the power of precipitating casein, which, after insolation, they lose. The coloring matter is therefore mixed with milk, and the exposed print immersed in it; the casein, and pigment with it, are precipitated on the protected parts, which form the blacks of the picture.

Early in 1863, Mr. Pouncy called attention to a new mode of carbon printing, which he had patented in the previous January. In the course of the summer the details were published,* and were found to embrace an important new principle in carbon printing, inasmuch as the picture was formed of a fatty ink, similar to that used in ordinary mechanical printing. Thin, transparent paper, like tracing paper, was coated with a mixture of carbon or other pigment, fatty matter, such as tallow or oil, bichromate of potash, or bitumen of Judæa, or both, and turpentine, or some equivalent body.† When

* "Photographic News," vol. vii, p. 169.

† It is worthy of note here that a very similar mixture appears to have been used at a very early date in obtaining a sensitive surface for photo-lithographic purposes. In January, 1863, Mr. A. Mactear read a communication to the Glasgow Photographic Association, describing a process of photo-lithography employed by Mr. Gibbons in 1859. The sensitive compound applied to the

dry, this is exposed under a negative, with the back in contact, according to the now recognized principle of obtaining half-tone. After exposure, the unaltered matter was removed by means of turpentine or other similar solvent, leaving an image with perfect gradation, in a material analogous to printing ink. The picture thus obtained was then mounted on white or tinted paper. The chief drawbacks to this process were: first, the impure lights which resulted from the tracing paper on which it was necessary to produce the prints to secure sufficient transparency in printing through the paper. This has, however, been overcome by adopting a transferring process in which the image is removed from tracing paper to any ground which may be chosen. The second difficulty was the long exposure, which was about three times as protracted as that necessary for silver printing on albumenized paper. Many of the results we have seen are, however, very excellent.

Later in the same year, Mr. Blair* expressed a conviction as to the disadvantages of any method of producing prints on waxed, oiled, or varnished paper, and proposed a method on plain paper, in which he endeavored to compensate for the tendency to lose half-tone when the prepared surface was presented direct to the light, by the mode of preparing the paper. It was at first coated with gelatine; then, when dry, with albumen and syrup (containing a little transparent color, to give a delicate half-tint to the paper). Subsequently the surface was coated with carbon powder, which was made to adhere by moistening the back of the paper. Finally, when required for

stone consisted of copal varnish, raw linseed oil, bichromate of potash, Brunswick black, mastic varnish, and turpentine, ground up together.

* "Photographic Notes," vol. viii.

use, it was floated, back downwards, on a solution of bichromate of potash. When dry, it was exposed with the face in contact with the negative. The washing away of unaltered sensitive material and color was effected by cold or warm water, and a brush, to which sometimes a little ammonia, or acetic acid, was added. Mr. Blair describes this method as giving, with care, pretty good results.

The next step in carbon printing marks an important epoch in its history: we refer to the introduction by Mr. Swan of a prepared tissue for producing the pictures, which permits exposure on one side, and washing away on the other. This step, together, with the complete system of operations connected with its use, made carbon printing practicable as a useful art. The process was first announced in the *Photographic News*, early in 1864, and during many succeeding months continued to occupy a large share of public attention. The results were as perfect as the most fastidious could desire, and the process was so simple in itself, and so clearly stated by Mr. Swan, that for the first time in the history of carbon printing, many experimentalists gave attention to the subject, and produced excellent pictures. As the mode of working is given in another chapter, it is not necessary to state it here.

In the course of the discussions elicited in the photographic journals by the publication of Mr. Swan's process, we learn that Mr. Davies, of Edinburgh, had, in the course of experiments in photo-lithography, produced transferred carbon prints as early as 1862; allusion to which was made in a paper read at the Edinburgh Society in February, 1863. A series of circumstances, however, prevented the publication of his process until July, 1864. He then described a method analogous to that already patented by Mr. Swan, namely, coating paper with gelatine, bichromate, and pigment; exposure with the prepared surface next the negative, mounting with a solution of shellac and Venice turpentine in alcohol, or with albumen, and then coagulating; soaking until the original paper leaves the gelatine and pigment, and then developing with hot water.

It is not necessary to mention all the minor modifications of Swan's method which were proposed; nor the various suggestions which grew out of the discussion of his process. We may, however, mention one or two of the latter. Mr. Frank Eliot suggested taking advantage of M. Poitevin's last process with perchloride of iron and tartaric acid. He proposed working with black paper, coating it with gelatine and white pigment, sensitizing with the iron salts and tartaric acid; then exposing, and developing, to obtain an image in white pigment on a black ground. If any attempt were made to carry out such a scheme, it is obvious that a pigment must be found which would not decompose the iron salt or be affected by it.

During the following year, Mr. M. Carey Lea published working details of two processes, both of which were analogous to some of the earlier processes. In the first,* the paper was prepared with a mixture of gelatine, glycerine, bichromate of potash, and water; and, after drying, was exposed under a positive cliché. The parts intended to form the lights become hardened and insoluble; the shadows, being protected from the light, swell and soften, but do not dissolve on immersing the print in cold water. The prints are left soaking, to remove as much of the color of the reduced chromic salt as possible from the lights; and afterwards, finely-powdered lampblack is applied, which adheres to the softened gelatine, and forms the picture. Mr. Lea points out that,

* "Philadelphia Photographer," vol. ii.

as in this process the lights are embodied in the shadows, and no part is washed away, there is no danger of losing the half-tones, as in some processes; and that he thinks it is possible to obtain some degree of gradation by this method.

The second process proposed by the same gentleman* is avowedly for the reproduction of subjects without half-tone. In this process he employs a mixture of gum-arabic, albumen, glycerine, double chromate of potash and ammonia, powdered graphite, and water. The mixture is of the thickness of honey, and is applied to the paper with a broad brush. When dry, it is exposed under a negative of any subject in line or stipple. After exposure, the right amount of which may be ascertained by examining the back of the print, it is developed by soaking in cold water, which removes the unaltered material, leaving clean lights and good blacks.

A more recent suggestion for a method of carbon printing was made in the course of last year by Dr. Gotschalk. He observes that graphitic acid, a substance prepared by the action of nitro-sulphuric acid upon graphite, is sensitive to light, which deoxidizes it, reducing it again to the condition of graphite. Hence it was proposed to prepare paper with a solution of this body, and expose it under a negative to the action of light. The first difficulty in the way is its insolubility, or sparing solubility, in any available menstruum. The trace, dissolved by water, applied to paper, is sufficient to color the paper brown when exposed to light. A method of checking its action, or fixing the print, would also be required. No practicable application of this substance for this purpose has yet been found.

Although Mr. Swan has perfected a process for us, by

* "Photographic News," vol. ix, p. 459.

USE
PEARL PAPER
FOR SALE BY ALL
STOCK DEALERS.

See what Prof. JOHN TOWLER says: "The surface of the paper is almost as structureless as glass. The albumen film does not blister in the toning and fixing, and it prints and tones with great ease and satisfaction; there is, too, a pearly whiteness of surface which is quite characteristic."

See what the EDITOR of THE PHILADELPHIA PHOTOGRAPHER says in speaking of Gurney's beautiful picture in the January number, page 31: "They (Gurney & Son), use the best of materials, employ the best workmen, &c., &c., and the prints are made on the celebrated PEARL PAPER. Negatives by a 4-4 Harrison Lens and an American Optical Co.'s Box."

See proceedings of American Photographic Society, held in New York, January 28, 1867, reported in the "Philadelphia Photographer," No. 38, page 57. "Humphrey's Journal," February 1, page 295. A committee reported the PEARL PAPER as giving the best results of any paper. See the following:

We take pleasure in recommending the PEARL PAPER as being superior to any that we have heretofore used.

J. GURNEY & SON, Broadway, New York.
R. A. LEWIS, Photographer and Manufacturer of the celebrated L. & H. Collodion, Chatham Street, New York.
A. JORDAN & Co. (successors to A. Bogardus), New York.
R. G. BARCLOW, Bowery, New York.
GEORGE F. PABST & Co., Bowery, New York.
FRED. ULRICH, Bowery, New York.
R. L. WOOD, Broadway, New York.

"We find the PEARL PAPER decidedly the best we have ever used."—A. S. WOOD, *Chatham Street;* A. S. TOPLANYI, *Manufacturing Photographer, and Compiler of Gen. Sherman's Illustrated Cumberland Campaign.*

BROOKLYN, March, 1868. "I have used over thirty reams of PEARL PAPER during the past year, and unhesitatingly pronounce it the best in the market."—JOSEPH HALL, *Manufacturing Photographer for the Trade.*

BROOKLYN, March, 1868. "I take pleasure in saying the PEARL PAPER is by far the best I can get anywhere."—A. H. SHEPHARD.

"We fully concur with the above."—JOHN C. CHAPMAN, *Excelsior Gallery, Brooklyn;* M. KIRTSON, *Newark, N. J.*

"I take pleasure in recommending PEARL PAPER to others, having used it myself with entire satisfaction."—A. RUNKEL, 130 *East Thirteenth Street, New York.*

Ask any *first-class operator* who has given the PEARL PAPER a fair trial, the universal answer is, ***"It is the Best."*** All Stockdealers sell it.

THE ZENTMAYER LENS,

For Views and Copying.

PATENTED MARCH, 1866.

These Lenses possess, pre-eminently, the following qualities:

Width of visual angle, ranging from 80° to 90°; depth of focus; extreme sharpness over the whole field; true perspective; freedom from all distortion in copying; portability and cheapness.

Each mounting is provided with a revolving Diaphragm, containing the stops of the different combinations for which they are designed. The larger ones are provided with an internal shutter for making and closing the exposure:

No. 1,	2½ in. focus,	3 x 3 plate,	$20 00
" 2,	3½ "	4 x 5 "	25 00
" 3,	5½ "	6½ x 8½ "	30 00
" 4,	8 "	10 x 12 "	42 00
" 5,	12 "	14 x 17 "	60 00
" 6,	18 "	20 x 24 "	90 00

No. 1 and No. 2 combined, . . .	$33 00
" 2 " 3 " . . .	40 00
" 3 " 4 " . . .	55 00
" 4 " 5 " . . .	75 00
" 5 " 6 " . . .	110 00
" 1, 2, and 3, " . . .	48 00
" 3, 4, and 5, " . . .	88 00

No. 3, with large mounting to combine with No. 4 and No. 5, $35.

No. 1 and No. 2, specially adapted for Stereoscopic Views, are furnished in matched pairs. No. 1, single, not to combine with other sizes, $36 a pair.

Lenses and mountings to form all six combinations, from 2½ to 18 inches, $173.

ZENTMAYER'S STEREOSCOPIC OUTFITS

Knowing that many photographers are retarded from making stereoscopic views on account of the high prices heretofore charged for lenses, thus depriving them of what might be a source of considerable revenue, I take pleasure in announcing that in order to obviate this, I offer a STEREOSCOPIC OUTFIT, consisting of

1 pair 3½ Focus Lenses, best quality, in mounting not to combine with other sizes.

1 American Optical Co.'s Stereoscopic Box, 4 x 7, made especially for these lenses.

1 Folding Tripod for box.

1 7 x 10 Tight Covered India-Rubber Bath and Dipper.

2 4 x 7 American Optical Co.'s Printing Frames.

The whole, except the tripod and printing frames, contained in a neat box, with handle, convenient for carriage, and weighing only 11 pounds.

Price, complete, $60.

JOSEPH ZENTMAYER, Manufacturer,

147 South Fourth Street, Philadelphia.

WILSON'S IMPROVED PHOTOGRAPHIC REST.

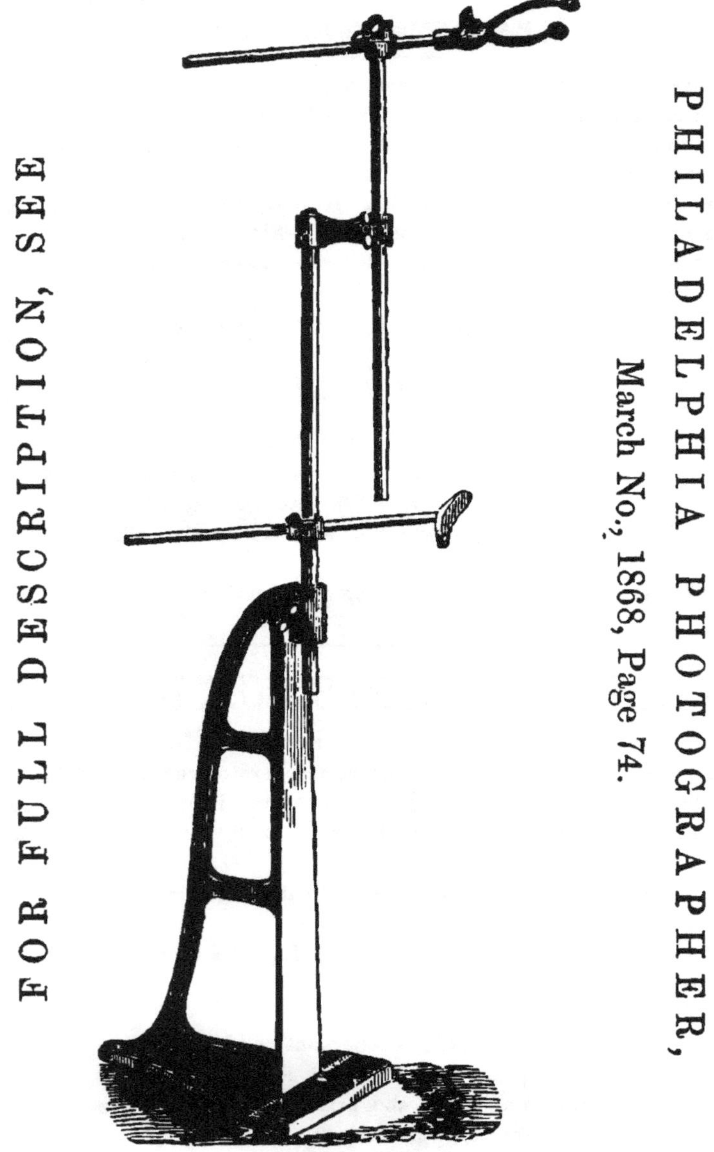

FOR SALE BY

Scovill Manufacturing Company,

4 BEEKMAN STREET, NEW YORK.

Wilson's Improved Photographic Rest,
AND POSING MACHINE.

For full description, see Philadelphia Photographer, April, 1868.

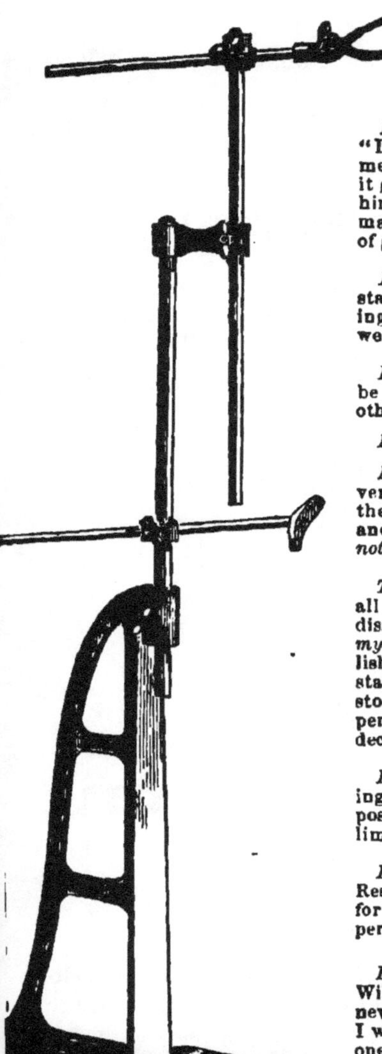

Highly Polished Rods, $25.00.
Unpolished " 18.00.

TESTIMONIALS.

From Charles G. Crane, 532 Arch Street, Philadelphia.—"I find it so valuable an addition to my posing arrangements that it now seems indispensable. The firm support it gives to the subject cannot but be an assurance both to him and to the operator, and the facility with which it is manipulated, and variety of ways it can be used, makes it of great value to the Photographer."

From F. S. Keeler.—"I use your 'Wilson Rest' constantly. It is worth a dozen of any other kind. For standing figures, especially, it is invaluable. The 'Rest' for hot weather."

From W. C. Westcott, Bridgeton, N. J.—"Mine could not be bought for double the price if I could not procure another."

From W. J. Baker, Buffalo, N. Y.—"I like it *much*."

Mr. J. H. Kent says of it: "I think it is excellent. It is very easily adjusted, and quite firm. I am sure it is just the thing that is needed. I made twenty-six 4-4 negatives and several cards with it one day, all with small stops, and *not one had to be retaken!* The *rest* kept the subjects easy."

To Messrs. E. & H. T. Anthony & Co., N. Y.—"It combines all the excellencies of the Sarony apparatus, with none of its disadvantages. *It paid for itself the first time it was used in my establishment.* A lady who had tried four different establishments in Paris, failed, as she declared, from inability to stand still. In a long exposure for a first-class picture, she stood like a statue with the 'Wilson Posing Machine.' A person of ordinary weight can absolutely recline in it. I decidedly prefer the polished ones."—G. G. Rockwood.

From C. D. Fredericks & Co., New York.—We are exceedingly pleased with the Wilson Posing Machine. We can pose our subjects by its aid in attitudes almost without limit, and they remain perfectly steady."

From W. Notman, Montreal.—"I have two of your new Rests, and candidly think they are a decided improvement for general use on anything before produced. They give a perfect support, and are easily handled."

From Jesse J. Groom, Philadelphia.—"I have been using Wilson's Posing Machine for two months, and can say I never made better work in my life than that made with it. I would not be without it under any circumstances. It is one of the best improvements of the photographic age."

From F. Gutekunst, Philadelphia.—"I use it, and look upon your Head-Rest as a decided improvement upon the old one."

FOR SALE BY ALL DEALERS.

WILSON, HOOD & CO., Manufacturers,

No. 626 (after July 1st, 822), Arch St., Philadelphia.

FERROTYPE PLATES,

EGG-SHELL AND GLOSSY.

PATENTED.

GRISWOLD FERROTYPE PLATE COMPANY,

PEEKSKILL, N. Y.,

SOLE MANUFACTURERS.

Sold by all Photographic Stockdealers in the United States and Canadas.

The attention of the Photographic Public is called to the low rates at which the Ferrotype Plates are now being sold. Their *Superiority* over every other similar article is *Universally Conceded;* and while the demand for them at former prices has exceeded the supply, efforts have for some time been directed toward manufacturing them in quantities sufficiently large to meet the pressing calls of Stockdealers and others promptly.

The *Standard of Excellence* which has secured the *High Reputation* of the Ferrotype Plate will be *Rigidly Maintained.*

CAUTION.

The GREAT POPULARITY which the EGG-SHELL brand of Ferrotype Plates has attained, has induced unscrupulous parties to assume that name for certain comparatively *worthless imitations. The liabilities of the vendors, as well as the manufacturers, of such plates, FOR A VIOLATION OF OUR TRADE MARK, IS UNQUESTIONABLE.*

☞ ***Get only Griswold's Genuine Egg-Shell Plates.***

All our plates are put up in strong, tight, neat, and handy boxes. The boxes are covered with brown Manilla paper, and labelled (in green and gold) on both sides, "Griswold's Ferrotype Plates;" on both ends, the number, size, and kind of plates.

EVERY PLATE GUARANTEED.

MOUNTFORT'S
SELF-DRYING
CRYSTAL VARNISH.

FOR

Melainotypes, Ambrotypes, and Negatives.

No Heat Necessary in Drying.

The only perfectly reliable Varnish in use, and which is supported and recommended by the most eminent Photographers in this Country.

It contains ingredients never before used in Varnishes, and possesses the following merits over every other brand:

It dries in a few seconds, WITHOUT HEAT, almost as hard as glass.

It is perfectly transparent, and does not diminish the intensity of the negative.

Negatives can be printed from in a few seconds after having been varnished.

It will not adhere to the paper, or become soft in printing.

It never chills if the plate is dry.

It is equally good for positives or negatives.

It is almost colorless, gives a beautiful gloss and brilliancy to Ambrotypes and Melainotypes; it is as hard as glass, and will not crack or peel.

It is the best Varnish made for Iron Plate Pictures.

It is used without heat, thereby saving the cost of burning alcohol.

Read what the Photographers and Stockdealers say of Mountfort's Crystal Varnish.

ABRAHAM BOGARDUS, of 363 Broadway, New York, says: "I have used a great many kinds of Photographic Varnish since I have been in the business (now over fifteen years), but never have I found anything to come up to Mountfort's Crystal Varnish. It is truly a big thing. It dries quickly, without the use of heat, is as smooth and as hard as the glass itself. It improves rather than diminishes the intensity of a negative, and it acts like a charm upon Porcelain, Iron Plates, and other small pictures. In fact, it is perfect in every respect, and I will use none other as long as I can get the true and genuine Mountfort's Crystal Varnish."

CHARLES H. WILLIAMSON, of 110 Fulton Street, Brooklyn, N. Y., says: "I do not know how we could well get along without Mountfort's Crystal Varnish. It has got to be as necessary to have it for preserving a good negative, as it is to have good lights and chemicals to produce a good negative; and it is so handy, requiring no heat in drying, thereby doing away with the old flaring Alcohol Lamp, and saves the cost of the Varnish by requiring no Alcohol. I consider it one of the greatest improvements, in the manufacturing departments of Photography, of the age.

JOSEPH THWAITES, of 691 Broadway, N. Y., says: "Please send me half a dozen bottles Crystal Varnish. I sent to a Stock House, as I had not time to send down to you, and they sent me a bottle of Varnish put up in some such style as Mountfort's Crystal Varnish. I paid no attention to it when it came into the Gallery, and my operator went on using it. It has ruined over five hundred dollars' worth of work. It is well that I saw the bottle as I did, or I would have been deeper into it. I have used your Crystal Varnish ever since you made it, and I have never had the least trouble with it, and I never want any other but yours. The man that will sell an imitation is as mean as the man that manufactures it."

I have in my possession as many as fifty more testimonials from the most prominent Photographers and Stockdealers in the country. Had I the space to print them, I would do so. Any person who wishes can see them at my office.

IMPROVED
CARTE ENVELOPES AND PICTURE STANDS.

Although hundreds of thousands of this capital receptacle for photographs have been sold, it is believed that many have not yet learned their value.

They are well made, of excellent paper, assorted tints, neatly gilt, and hold one carte or one dozen.

They are the

SAFEST ENVELOPES FOR MAILING.

The most beautiful to deliver pictures in.

When the flap is folded back (see cut), they form a beautiful stand for the picture.

The artist's name may be printed on them, thus circulating his advertisement, and keeping his business before the people.

Every photographer and ferrotypist who prides himself on neatly delivering his work, should use these envelopes. Two sizes (carte and cabinet), and two shapes (arch-top and oval), are made. All colors. Put up in neat paper boxes.

For Sale by all Dealers.

Pure Chemicals.

Photographers who wish to use reliable Chemicals, should purchase

THE SCALE

TRADE MARK.

CHEMICALS.

A FULL LIST OF CHEMICALS USED IN

PHOTOGRAPHY AND THE CARBON PROCESS,

Guaranteed pure and reliable.

Challenge Varnish.

"THE BEST"

In use in most of the leading New York Galleries, including SARONY'S, GURNEY'S, KURTZ'S, and others.

6 oz. Bottles, 50 cts.; Quarts, $2.25; Half Gallon, $4.25.

Mardock's Collodion.

"UNSURPASSED"

For rapidity, cleanliness in working, and exquisite detail.

$1.75 per pound.

CATALOGUES SENT TO ANY ADDRESS.

MARDOCK & WALLACE,

Photographic Chemists, 417 Broome Street,

NEW YORK.

ROME ALUM & BICHROMATE OF POTASH

Specially prepared for the Carbon Process.

L. MARTIN & COMPANY,

MANUFACTURERS OF

PURE CHEMICALS,

NEW YORK AND PHILADELPHIA.

Sold by Dealers in Photographic Stock, Drugs, &c.

L. M. & Co. take great pleasure in presenting to PHOTOGRAPHERS the following testimonials in favor of their

Pure Nitrate of Silver, Acetic, and other C. P. Acids, &c.

TESTIMONIALS.

From Messrs. WENDEROTH, TAYLOR & BROWN, *Artists, Philadelphia and New York, well known as gentlemen of great scientific attainments in their profession.*

February 28, 1868.

MESSRS. L. MARTIN & CO., Manufacturing Chemists, &c.

Gentlemen: It affords us much pleasure to recommend to Photographers, &c., chemicals such as we have been using from your establishment during the last year, viz., Nitrate of Silver and Acetic Acid, which have not only been found the best, but the most reliable we ever used.

Yours truly,

WENDEROTH, TAYLOR & BROWN.

From JOHN P. SOULE, ESQ., *the very extensive and well known Publisher of Photographs,* 199 *Washington St.*

BOSTON, March 9, 1868.

MESSRS. L. MARTIN & CO.—Gents: I have used your Nitrate of Silver for over a year, with uniform good success, and it gives me great pleasure to recommend it to Photographers and others, who desire a pure article.

Yours, truly,

JOHN P. SOULE.

First Premium Awarded by the American Institute of N. Y., Oct., 1867, *over all competitors, as follows:*

"L. Martin & Co., for Chemicals, especially for Purity of Acids, Iron by Hydrogen, Nitrate of Silver, and Salts of Bismuth.—1st Premium."

Extract from Rep. of Committee American Pharmaceutical Association, 1867.

"Acetic acid. This acid has the sp. gr. 1.048, is entirely free from any empyreumatic odor, and is the best acid of American manufacture your Committee have seen."

Extract from Letter of Prof. JOHN M. MAISCH, *Philadelphia College of Pharmacy, a gentleman of world-wide reputation.*

PHILADELPHIA, Sept. 20, 1866.

MESSRS. L. MARTIN & CO.: "It gives me great pleasure to state that I have examined some of your chemicals, among them Sulphate of Morphia, Nitrate of Silver, and Acids, and found them to be in every respect satisfactory for purity as well as appearance."

Pure Nitrate of Silver, Chloride of Gold, Acet

And other pure Acids.

Vogel's Photometer.

Now Ready.

SENT FREE TO ANY ADDRESS FOR $2.50.

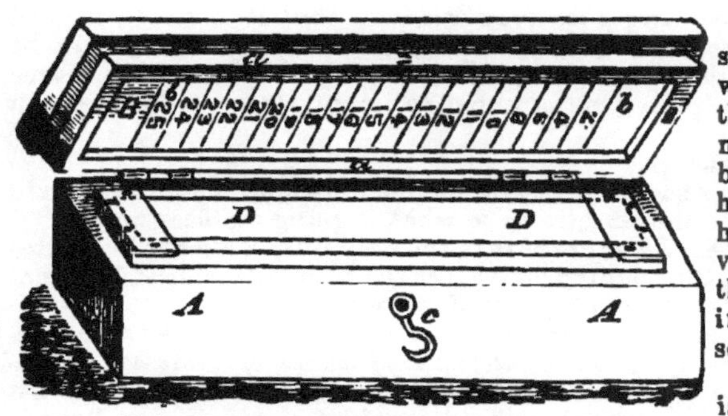

Every glass-room should be furnished with them for timing the exposure of the negative. Every Carbon Printer should have them for timing his prints. Every Silver Printer should use them. Fully described in this Manual. Please see.

Vogel's Photometer is the only reliable means of obtaining the proper time *certainly* for exposing carbon prints. Vogel's Photometer is used by Mr. Swan, Patentee of the Carbon Process. Send for one.

WILSON, HOOD & CO., Manufacturers,

For sale by all Dealers. **626 (after 1st July, 822) Arch St., Philada.**

Wenderoth, Taylor & Brown,

Artists & Photographers

No. 914 Chestnut St., Philada.

PHOTO-MINIATURES, IMPROVED IVORYTYPES, OPALOTYPES ON PORCELAIN, PHOTOGRAPHS OF ALL SIZES AND STYLES, PLAIN AND PAINTED IN THE FINEST MANNER, BY GENUINE ARTISTS.

ALSO, JUST INTRODUCED, A FINE LINE OF PRINTS, ENGRAVINGS, CHROMO-LITHOGRAPHS, &c., &c.

Agents for superior Artists' Materials.

Please see other advertisement.

www.ingramcontent.com/pod-product-compliance
Lightning Source LLC
Chambersburg PA
CBHW021938190326
41519CB00009B/1061